AF323029

CORROSION IN SEAWATER SYSTEMS

ELLIS HORWOOD SERIES IN CORROSION AND ITS PREVENTION

Series Editors:
A. D. MERCER, formerly of the National Physical Laboratory;
D. R. HOLMES, Manager/Consultant, High Temperature Corrosion Projects, National Corrosion Service, National Physical Laboratory

This series provides up-to-date compelling texts on the theory and practice of corrosion science and technology. There is special emphasis on cost-effective preventive measures and on the most recent developments in new materials and techniques for the study of corrosion products. The books will usually be at the postgraduate research or practising corrosion engineer/technologist level, but will also be suitable for informing advanced undergraduate and master courses of the most recent developments. Together, they will provide substantial coverage of the latest developments in corrosion phenomena and protective measures.

PROTECTION OF METALS FROM CORROSION IN STORAGE AND TRANSIT
P. D. DONOVAN, Principal Scientific Officer, Ministry of Defence, Royal Armament Research and Development Establishment
CORROSION IN SEAWATER SYSTEMS
Editor: A. D. MERCER, formerly of the National Physical Laboratory
BASIC CORROSION AND OXIDATION, Second Edition
J. M. WEST, Department of Metallurgy, University of Sheffield
CORROSION OF REINFORCEMENT IN CONCRETE CONSTRUCTION
Editor: A. P. CRANE, Pall Process Filtration, Hampshire
CORROSION AND MARINE GROWTH ON OFFSHORE STRUCTURES
Editors: J. R. LEWIS, Senior Lecturer in Chemistry, University of Aberdeen, and A. D. MERCER, Principal Scientific Officer, National Physical Laboratory, Teddington.
CORROSION DATA FROM POLARIZATION MEASUREMENTS
R. STEFEC, Prague Institute of Chemistry and Technology, Czechoslovakia

Books in Corrosion Science Published for the Institution of Corrosion Science and Technology

CATHODIC PROTECTION: Theory and Practice
Editors: V. ASHWORTH, Global Corrosion Consultants, Telford, and C. J. L. BOOKER, City of London Polytechnic
DEWPOINT CORROSION
Editor: D. R. HOLMES, Manager/Consultant, High Temperature Corrosion Projects, National Corrosion Service, National Physical Laboratory
COATINGS AND SURFACE TREATMENT FOR CORROSION AND WEAR RESISTANCE
Editors: K. N. STRAFFORD and P. K. DATTA, School of Materials Engineering, Newcastle upon Tyne Polytechnic, and C. G. GOOGAN, Global Corrosion Consultants Limited, Telford
PLANT CORROSION: Prediction of Materials Performance
Editors: J. E. STRUTT and J. R. NICHOLLS, School of Industrial Science, Cranfield Institute of Technology, Bedford

CORROSION IN SEAWATER SYSTEMS

A. D. MERCER
formerly of the National Physical Laboratory

ELLIS HORWOOD
NEW YORK   LONDON   TORONTO   SYDNEY   TOKYO   SINGAPORE

First published in 1990 by
ELLIS HORWOOD LIMITED
Market Cross House, Cooper Street,
Chichester, West Sussex, PO19 1EB, England

A division of
Simon & Schuster International Group
A Paramount Communications Company

© Ellis Horwood Limited, 1990

All rights reserved. No part of this publication may be reproduced, stored in a retrieval
system, or transmitted, in any form, or by any means, electronic, mechanical, photocopying,
recording or otherwise, without the prior permission, in writing, of the publisher

Typeset by Ellis Horwood Limited
Printed and bound in Great Britain
by Hartnolls, Bodmin, Cornwall

British Library Cataloguing in Publication Data

Mercer, A. D. (Antony Donald) 1928–
Corrosion in seawater systems.
1. Corrosion by seawater
I. Title
620.11223
ISBN 0–13–388703–0

Library of Congress Cataloging-in-Publication Data

Corrosion in seawater systems / [edited by] A. D. Mercer
p. cm. — (Ellis Horwood series in corrosion and its prevention)
ISBN 0–13–388703–0
1. Sea-water corrosion. 2. Stainless steel — Corrosion. 3. Corrosion and anti-corrosives.
I. Mercer, A. D. (Antony Donald), 1928– . II. Series.
TA462.C65543   1990
620.1′1223–dc20

90–42666
CIP

Table of contents

Table of contents

Foreword

This book is based on the material presented at two symposia organized by the Materials Preservation Group of the Society of Chemical Industry. Although the main subjects of the symposia were respectively the corrosion behaviour of stainless steels and corrosion aspects of chlorination in seawater, there were a number of related topics considered, including marine fouling and the performance of other materials, notably copper-based alloys. It was therefore appropriate to bring both collections together in a single volume under the general title of *Corrosion in Seawater Systems*. All the chapters have been reviewed by the authors since the original presentations and revised, as necessary, for publication.

The chapters are grouped into three Parts concerned with (1) stainless steels, (2) chlorination and (3) applications. There is inevitably some overlap in subject matter between the Parts but, nevertheless, these three headings represent the main themes of the chapters in the respective Parts.

Thus, Part I opens with a chapter on new developments in stainless steels for resistance in seawater and provides analysis and sources of these materials. This is followed by a chapter concerned exclusively with the cast forms of stainless steels and then a consideration of the welding of stainless steels in relation to corrosion behaviour. The Part concludes with a paper on the galvanic compatability of high alloy stainless steels in seawater.

Part II opens with a detailed account of marine fouling and the need for, and application of, chlorination treatments. The performance of copper-based alloys as corrosion-resistant materials in chlorinated waters is then considered followed by that of the resistance of stainless steels in such media.

Part III has papers dealing with specific aspects of the principles and properties discussed in Parts I and II. The Part opens with a consideration of corrosion and chlorination in offshore seawater systems and continues with the specific topic of condenser tubes in electrical generation plant. The Part continues with two chapters on desalination systems with examinations of the use of high alloy stainless steels and of corrosion problems caused by chlorination in this technology of fresh water production and concludes with a description of electrochlorination systems and materials used in there.

J. W. Oldfield
Chairman
Materials Preservation Group
Society of Chemical Industry

Part I
Stainless steels

1

New developments in stainless steels for resistance to corrosion in seawater

G. Tither†
Climax Molybdenum Co. Ltd, 29 Gresham Street, London EC2V 7DA

INTRODUCTION

The objective of this short introductory chapter is to present a brief outline of the nature and properties of the so-called seawater-resistant grades of stainless steel as a background to the more specific chapters to be presented in this volume. General corrosion is not normally a problem in seawater and data to be presented will concentrate on pitting, crevice and stress corrosion cracking. After a general discussion of the ferritic, austenitic and duplex grades, this chapter will concentrate on the crevice corrosion performance of the stainless steels under laboratory conditions.

It is probably correct to state that the introduction of more sophisticated steel processing techniques such as AOD (argon oxygen decarburization) and VOD (vacuum oxygen decarburization), which enable much lower levels of carbon and nitrogen to be obtained, is the base for 'new' developments over the past decade. While the VOD process produces C+N levels of around 100 ppm, the AOD process results in somewhat higher levels of up to 400 ppm. In the latter case the ferritic steels have to be stabilized so that precipitation of chromium-rich carbides does not occur. This precipitation is detrimental in that it produces a chromium-depleted zone which if continuous is susceptible to intergranular corrosion as well as pitting corrosion. The problem is particularly prominent in the heat-affected zones (HAZs) of weldments. Dundas and Bond [1] have defined the minimum addition of Nb or Ti necessary to produce stabilization of 18 Cr 2 Mo alloys containing 0.2–0.5 per cent (C+N) as:

$$Nb \text{ or } Ti = 0.20\% + 4(C+N) \tag{1.1}$$

† Now with Niobium Products Company Inc., 300 Corporate Center Drive, Pittsburgh, PA 15108, USA

FERRITIC STAINLESS STEELS

In ferritic stainless steels both chromium and molybdenum are essential elements for good corrosion resistance. Many attempts have been made to quantify their effect and it is now common to see a pitting resistance equivalent (PRE) referred to:

$$PRE = Cr\ wt\% + (3.3 \times Mo\ wt\%) \tag{1.2}$$

Since a fully ferritic microstructure is necessary for good resistance to chloride attack, the upper limit of alloy addition is in the region of 29 per cent Cr and 4 per cent Mo. A 29 per cent Cr–4 per cent Mo steel is demonstrably one of the most resistant ferrous alloys to corrosion attack in aqueous chloride solutions. Fig. 1.1

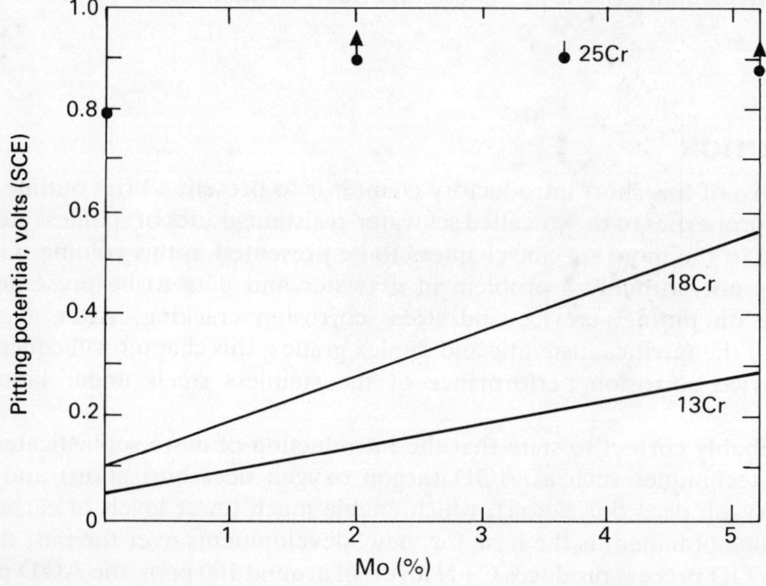

Fig. 1.1 — Pitting potentials for high purity Fe–Cr alloys in 1 N NaCl at 25°C [2].

illustrates the beneficial effect of molybdenum on the pitting potential of ferritic steels at several chromium levels [2].

Ferritic stainless steels are subject to recrystallization and possible grain growth in the HAZs of weldments and this tends to limit the commercial availability of flat products to sheet of 3 mm thickness or less if these are to be used in fabrications. Although niobium stabilization is helpful in increasing the recrystallization temperature and minimizing grain growth, it is not the total answer to the problem. Further benefit can be obtained by the addition of nickel. While nickel has a slight advantageous effect on corrosion resistance of ferritic stainless steels, its main attribute is to give improved toughness by lowering the impact transition temperature, Fig. 1.2 [2]. This improvement in toughness afforded by nickel allows thicker plate product to be used successfully.

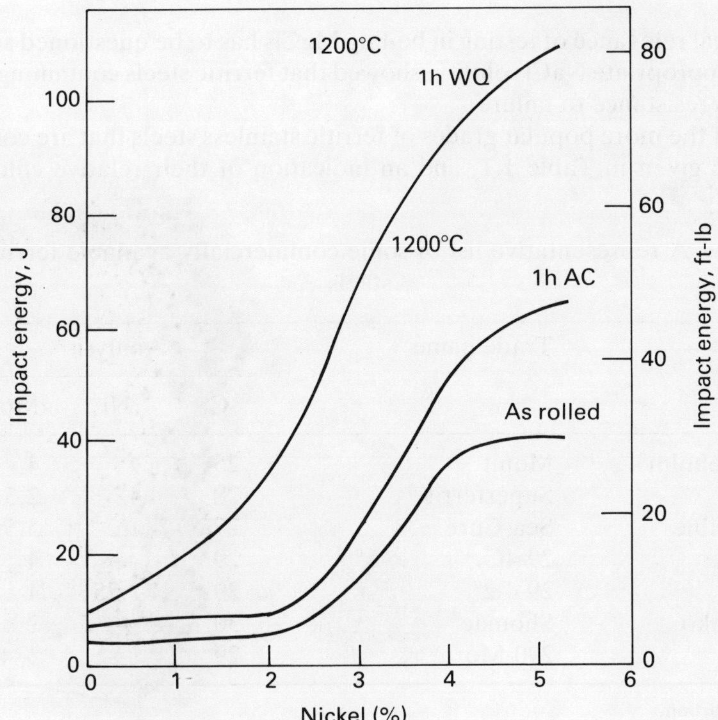

Fig. 1.2 — The effect of nickel on the toughness of 25 Cr–0.7 Nb–3 Mo–Ni ferritic stainless steels at 25°C [2].

The amount of nickel that may be added is restricted by the same limitations that apply to molybdenum, that is that the resultant microstructure must be fully ferritic and free from secondary phases otherwise there is a marked deterioration in corrosion resistance. At the 25 per cent Cr level, which is common for a number of seawater stainless steel grades, the maximum nickel content is in the region of 4 per cent. At a lower Cr level, say 18 per cent Cr, then only about 1 per cent Ni can be tolerated (Schaeffler diagram). The range over which the ferrite phase is stable is a balance between the ferrite-forming elements Cr, Mo and Si and the austenite-forming elements Ni, Mn, N. C, Cu, etc.

The simple FeCrMo alloys are highly resistant to stress corrosion attack and may effectively be considered immune for all practical purposes. The addition of nickel, copper and cobalt decreases the resistance to failure [3]. The compositions of annealed FeCr alloys that are immune to cracking in boiling $MgCl_2$ are roughly described by:

$$Ni\ wt\% + (3 \times Cu\ wt\%) < 0.9 \tag{1.3}$$

The practical relevance of testing in boiling $MgCl_2$ has to be questioned since testing in a more appropriate NaCl solution showed that ferritic steels containing nickel had a very high resistance to failure.

A list of the more popular grades of ferritic stainless steels that are commercially available is given in Table 1.1, and an indication of their relative chloride stress

Table 1.1 — A representative list of some commercially available ferritic stainless steels

Producers	Trade name	Analysis			
		Cr	Ni	Mo	
Nyby-Uddeholm	Monit	25	4	4	Ti
TEW	Superferrit	28	4	2.5	Nb
Trent/Crucible	Sea-Cure	27.5	1.2	3.5	Ti
Allegheny	29-4C	29	—	4	Ti
Allegheny	29-4-2[a]	29	2.25	4	—
Showa Denko	Shomac[a]	30	—	2	
Usinor	290-Mo	29	—	3–4	Ti

[a]Ultra Low Carbon.

corrosion-cracking resistance is shown in Table 1.2.

Table 1.2 — Test to indicate how the seawater stainlesses withstand chloride stress-corrosion cracking. (Private communication, Amax Materials Research Center, Ann Arbor, MI)

Stainless steel	Stress corrosion cracking test		
	Boiling 42% $MgCl_2$	Wick test	Boiling NaCl, pH=7
Sea-Cure	F	P	P
AL 29-4-2	F	P	P
AL-29-4C	P	P	P
Monit	F	P	P

P=Passed.
F=Failed.

AUSTENITIC STAINLESS STEELS

In many ways the austenitic stainless steels are much more easy to handle than the ferritic grades, and they do not experience the same limitations on thickness. The resistance to chloride attack is again primarily a function of chromium and molybdenum with pitting resistance equivalent equally applicable to both ferritic and austenitic grades. However, in contrast to ferritic grades where nitrogen is very detrimental, nitrogen in austenitic grades has been found to be extremely useful. It acts both as an austenite stabilizer and an austenite strengthener and also improves the resistance to chloride attack. Alloying with nitrogen retards sigma-phase formation and allows for the production of thicker sections.

The pitting resistance formula has been modified to include nitrogen such that:

$$PRE = Cr\ wt\% + (3.3 \times Mo\ wt\%) + (16\ to\ 30 \times N\ wt\%) \tag{1.4}$$

Nitrogen additions in the region of 0.10–0.20 per cent or more are now becoming common practice in seawater-resistant stainless steels.

Because of the need to maintain a fully austenitic structure free from embrittling phases such as chi and sigma, the chromium content of the earlier group of austenitic stainless steels was typically in the range of 18–20 per cent Cr. A molybdenum level in excess of 6% is necessary in order to generate an adequate resistance to attack under the most severe crevice conditions in fully aerated ambient seawater. The newer steels such as Sanicro 28, 20Mo-6, etc. (see Table 1.3) contain up to 27 per cent Cr, at which level a lower molybdenum content can be tolerated. This compares with the need for 25–26 per cent Cr and $3\frac{1}{2}$ per cent Mo for a similar performance by the ferritic steels.

Nickel is present in amounts up to around 35 per cent in these steels primarily as a means of controlling and maintaining an austenitic microstructure. Very high nickel contents are beneficial with respect to increasing the resistance to SCC and help to minimize the active corrosion rate by assisting in repassivation.

As a group the austenitics are susceptible to stress corrosion cracking (SCC), although this is dependent on the nickel content. Both nickel-free ferritic steels and high nickel austenitic grades are resistant [4] while maximum susceptibility to failure in boiling 42 per cent $MgCl_2$ solution occurs around 8–10 per cent Ni. The practicality of testing in such a solution raises questions and more recent work [5] has examined the effect of nickel content on the SCC of FeNiCr alloys in the standard NaCl solution at 105°C. The effect of nickel content on the SCC threshold stress intensity, K_{ISCC}, for alloys with about 18 per cent Cr is shown in Fig. 1.3. In this case a maximum susceptibility also exists but is shifted to a higher nickel range of 10–25 per cent Ni. Speidel [5] also evaluated the effect of molybdenum under similar circumstances (Fig. 1.4) and found a very strong beneficial effect on K_{ISCC} as the molybdenum content was increased from 0 to 5 per cent Mo. From a practical viewpoint, failure is unlikely to occur at temperatures below 60°C and SCC problems can be disregarded in some major applications such as seawater-cooled power station condensers. In general, the new high alloy austenitic seawater grades show a better resistance to stress corrosion cracking than the conventional type 304 (18 Cr–8 Ni)

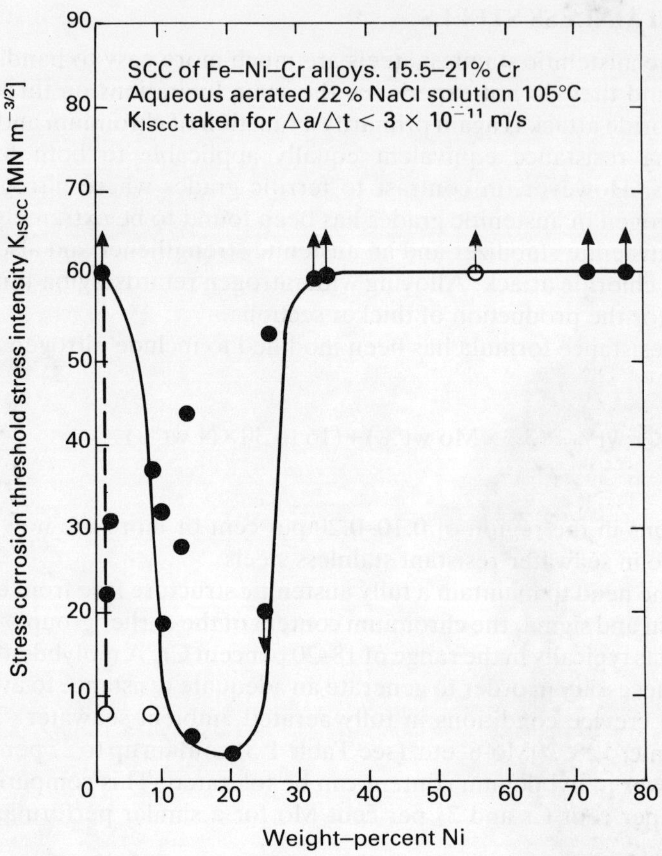

Fig. 1.3 — Effect of nickel content on the SCC resistance, K_{ISCC}, of Fe–Ni–Cr alloys with about 18 per cent Cr [5]. The data in this figure were derived from commercial stainless steel.

and type 316 (18 Cr–12 Ni–3 Mo) steels. A representative listing of commercially available austenitic steels is given in Table 1.3.

DUPLEX STAINLESS STEELS

Duplex stainless steels have probably not yet achieved their full potential in seawater application and tend to be somewhat underrated. They have some very positive advantages in relation to strength and a capability of being produced in moderately thick wrought products or in castings. The aim is to obtain a 50 per cent ferritic/50 per cent austenitic, sigma-free (or other deleterious constituents) microstructure. This is achieved by balancing the ferrite and austenite-forming elements using the Schaeffler diagram as an initial guide. The balance of microstructural constituents can be varied over a wide range by variation in solution treatment temperature (sigma and other precipitates may also be formed if the cooling rate is too slow) and

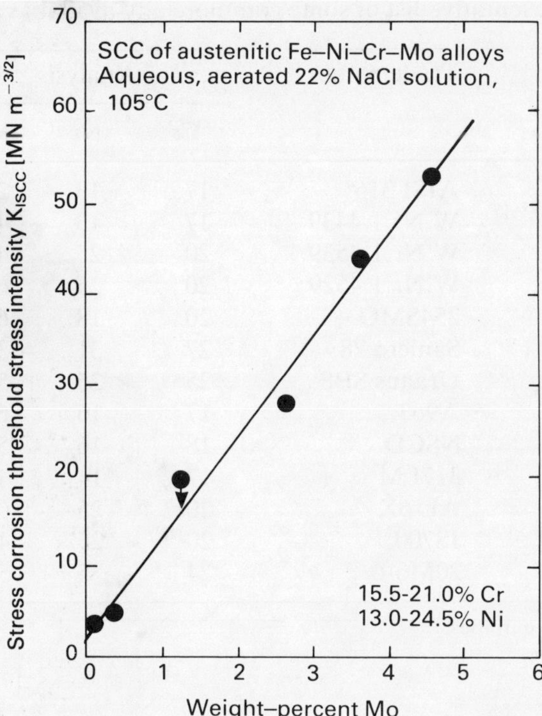

Fig. 1.4 — Effect of molybdenum content on the SCC resistance K_{ISCC}, of austenitic stainless steels [5].

consequently heat treatment practice and welding conditions need to be optimized if maximum performance is to be attained.

The one thing that distinguishes the new generation of duplex stainless steels is the use of nitrogen additions, usually within the range of 0.1–0.25 per cent N. Nitrogen is a well known austenite former and strengthener which consequently tends to minimize sigma formation. Additionally, in the duplex steels nitrogen tends to minimize segregation and delays the onset of carbide precipitation. The overall and important effect is that nitrogen greatly improves resistance to crevice attack in chloride solutions. Indeed, it is considered by some that nitrogen is essential in duplex stainless steels which have to be welded without subsequent heat treatment in that it maintains adequate corrosion resistance in the heat-affected zone.

Corrosion resistance in chloride environments is again a function of the chromium, molybdenum and nitrogen contents and the same PRE formula is applicable.

The duplex steels are not immune to stress corrosion but the resistance to failure is fairly high and consistent with the relatively low nickel content. A degree of 475°C embrittlement can occur and use may be restricted to temperatures below 300°C. As in the case of the other types of stainless steel, the corrosion performance of the duplex grades will be dealt with in the next section. A typical list of commercially available grades is given in Table 1.4.

Table 1.3 — A representative list of some commercially available austenitic steels

Producers	Trade name	Analysis				
		Cr	Ni	Mo		
General	AISI 316[a]	17.5	13	2.5	—	
Several European	W.Nr.1.4439	17	13	4.5	N	
Several European	W.Nr.1.4539	20	25	4.5	Cu	
VDM	W.Nr.1.4529	20	25	6	Cu	N
Avesta	254SMO	20	18	6.1	Cu	N
Sandvik	Sanicro 28	27	31	3.5	Cu	
Creusot Loire	Uranus SB8	25	25	5	Cu	N
VEW	A963	17	16	6.2	Cu	N
Ugine	NSCD	18	16	5.2	Cu	
Several US+BSC	317LM	17	14	4.5	—	
Allegheny	AL-6X	20	25	6	—	
Jessop	JS700	20	25	4.5	Nb	
Carpenter	20Mo-6	24	35	5.8	Cu	

[a]Included for comparative purposes.

Table 1.4 — A representative list of some commercially available duplex stainless steels

Producers	Trade name	Analysis				
		Cr	Ni	Mo		
Creusot Loire	Uranus 50	21	7	2.5	Cu	
Several European	W.Nr.1.4462	22	5	3	N	
VEW	A905	25	4	2.3	Mn	N
Nyby-Uddeholm	44LN	25	6	1.7	N	
Carpenter	7Mo	25	4	1.5	—	
Langley Alloys	Ferralium	25	5	3	Cu	N
Nippon Stainless	DP-3	25	7	3.1	Cu	W
Mather & Platt	Zeron 25	25	7	2	N	

CORROSION DATA

The various types of stainless steels have been subjected to extensive crevice corrosion testing by Bond and Dundas [6]. The necessary minimum alloying additions (Cr, Mo, Ni, N) to provide good corrosion resistance in seawater have been defined and give excellent pointers to the requirements of steel composition for seawater applications.

The implication of a seawater-resistant grade is a material suitable for prolonged service in raw fully aerated seawater at ambient temperature under low velocity conditions. Most of their evaluation was directed towards the conditions experienced in seawater-cooled power station condensers where the most critical factor is the resistance to crevice corrosion attack.

In the seawater exposure test serrated washer type specimens were used. Multiple crevice washers having 20 grooves and plateaux were fixed to the specimens and tightened to a specific torque. Three specimens of each steel were exposed initially in ambient fully aerated seawater flowing at a velocity of 0.6 m/s for two years. This provided a total of 120 crevices permitting a degree of statistical assessment. In a second method, a series of tests was conducted at 25°C in fresh filtered seawater at a velocity of <0.1 m/s.

The ferritic stainless steels were resistant to corrosion at Mo contents of at least 3 per cent Mo, while the austenitic steels, with a wider range of Cr and Mo content, exhibited behaviour from complete resistance to severe attack (Table 1.5). The superior performance of the 20 Cr–, 6.1 Mo–, 18 Ni steel (S31254) relative to the 20 Cr–, 6.5 Mo–, 25 Ni steel (N08366) was most probably due to the higher nitrogen content of the former, 0.22 per cent N, compared with 0.02 per cent N.

Bond and Dundas [6] also showed the importance of combined Cr and Mo content in determining the resistance of ferritic stainless steels to seawater. Steels with at least 25 per cent Cr and 3 per cent Mo were mainly completely unattacked or suffered only superficial (less than 0.05 mm) penetration at 1 or 2 of the possible 120 crevice sites.

The evaluation of the austenitic stainless steels was undertaken in two classes, those with low nitrogen content (0.02–0.08 per cent) and those with high nitrogen content (0.19–0.33 per cent). In contrast to the highly alloyed ferritic steels, the highly alloyed, but low N, austentic steels underwent significant attack (Table 1.6). The obvious indication is that in low N austenitics, higher Cr and Mo levels may be necessary to produce reasonable corrosion resistance.

Increasing the nitrogen content to about 0.2 per cent produced a substantial increase in corrosion resistance. The effect of Cr and Mo in these high N steels was substantial and indicated that a Cr content of 25 per cent and a Mo content of 3.5–4.6 per cent Mo, similar to that required for ferritic stainless steels, provides a high degree of resistance for high N (0.20 per cent) austenitic stainless steel.

The crevice corrosion behaviour of a series of duplex stainless steels is presented in Table 1.7. Only the two steels with the highest Cr, Mo and N contents showed appreciable corrosion resistance. These two steels also had the greatest resistance to crevice corrosion in 10 per cent $FeCl_3$.

The corrosion behaviour of duplex steels is complicated by the partitioning of alloying elements between the austenite and ferrite; the ferrite is enriched in Cr and Mo, and the austenite is enriched in Ni and N. In the present steels Bond and Dundas [6] found that the Cr partition ratio (Cr in ferrite/Cr in austenite) was reduced from 1.4 to 1.2 as the N content increased. Hence, nitrogen plays a dual role in improving the corrosion resistance of duplex steels. First, it reduces the partitioning of Cr, maintaining a higher Cr content in the austenite. Second, as shown for the fully

Table 1.5 — Results of crevice corrosion tests in ambient seawater, filtered seawater at 25°C [6]

Nominal steel comp (wt%)				Fraction of creviced area attack (%)				Maximum penetration (mm)			
				Ambient seawater				Ambient seawater			
							Filtered				Filtered
Cr	Mo	Ni	UNS No.	2 ms[a]	9 ms[b]	24 ms[a]	2 ms	2 ms[a]	9 ms[b]	24 ms[b]	2 ms
26	1.0	0.2	S44626	1	—	—	1.7	0	—	—	0.38
20	3.0	—	—	0	0	0	21.6	0	0	0	0.30
25	3.5	—	—	0	0	1.7	0	0	0	0.2	0
25	3.5	2.0	—	—	0	0	0	—	0	0	0
25	3.5	4.0	—	0	0	0	0	0	0	0	0
29	4.0		S44700	0	0	0	0	0	0	0	0
29	4.0	2.0	S44800	1.7	0	0	0	<0.02	0	0	0
18	2.0	11	S31603	7.2	—	—	100	0.23	—	—	1.41
20	4.5	25	N08700	0	5	5	5.8	0	0.09	0.11	1.49
19	4.3	25	N08904	0	1.7	0.8	99.2	0	0.06	0.05(0.08)[c]	0.83
20	6.5	25	N08366	0	0.8	0.8	0.8	0	0.01, (0.02)[c]	0.02	0.23
20	6.0	18	S31254	—	0	0	0	0	0	0	0

[a]The temperature range was 15.6–25°C.
[b]The temperature range was 5–31°C.
[c]Attack at fouling sites.

Table 1.6 — Crevice corrosion behaviour of austenitic stainless steels [6]

Nominal steel comp (wt%)				Ressitance in filtered seawater at 25°C (77°F)		Critical crevice corrosion temperature (°C) in 10% $FeCl_3.6H_2O$ (pH 1) (maximum) temperature of no corrosion
				Creviced area attacked (%)	Maximum penetration (mm)	
Cr	Mo	Ni	Site			
Low nitrogen steels (0.02 to 0.08%)						
18	2	11	2	99.2	>1.6	−2.5
			1	100	1.41	—
19	4	25	2	99.3	1.1	0
			1	99.2	0.83	—
20.5	4.5	25	2	10.0	0.80	15
			1	5.8	1.49	—
20	4.5	25	1	39.3	0.73	2.5
20	6.5	25	2	6.7	0.50	17.5
			1	0.8	0.23	—
27	3.5	31	2	3.3	0.34	12.5
22	5	26	2	13.3	0.20	22.5
High nitrogen steels (0.19 to 0.33%)						
25	2	22	2	3.4	0.75	10
25	3	17	2	18.3	0.42	27.5
24	3.5	22	2	2.5	0.21	15
20.5	6	18	2	1.7	0.04	32.5
				no attack	—	—
25.5	3.5	22	2	0.8	0.02	25
25	4.5	22	2	no attack	—	35

Table 1.7 — Crevice corrosion behaviour of duplex stainless steels [6]

Nominal steel comp (wt%)			Resistance in filtered seawater at 25°C (77°F)		Critical crevice corrosion temperature (°C) in 10% $FeCl_3.6H_2O$ (pH 1) (maximum) temperature of no corrosion
Cr	Mo	Ni	creviced area attacked (%)	Maximum penetration (mm)	
25.5	1.9	11	13.3	0.98	2.5
25	3	10	54.0	0.44	10
22	3	5.7	8.3	0.40	17.5
25	3	10	4.1	0.21	0
26	3.3	5.5	5.0	0.05	22.5
25.5	3.3	5.5	0.8	0.03	40

austenitic steels, nitrogen itself improves the corrosion resistance of the austenite phase.

SUMMARY

In summary, to achieve a high level of resistance to SCC in seawater grades the choice of steel should be either (a) a fully ferritic stainless steel, or (b) a high alloyed austenitic steel with a Mo content of at least 6 per cent.

To minimize crevice corrosion attack in seawater, ferritic, austenitic and duplex stainless steels should contain at least 25 per cent Cr and 3.5 per cent Mo. In addition, the austenitic and duplex grades also require a high nitrogen level of around 0.20 per cent. Further, lower chromium steels (18–20 per cent) need at least 6 per cent Mo to provide good corrosion resistance.

REFERENCES

[1] Dundas, H. J. and Bond, A. P. Intergranular corrosion of stainless alloys, *ASTM STP 656* (1977), p. 162.
[2] Steigerwald, R. F. *et al. Corrosion*, **33** (8), 279–295, August 1977.
[3] Bond, A. P. and Dundas, H. J. *Corrosion*, **24**, 344–351, 1968.
[4] Copson, H. R. *Physical Metallurgy of Stress Corrosion Fracture*, p. 247, Interscience, New York, 1959.
[5] Speidel, Marcus O. *Metallurgical Transactions A*, **12A** (5), May 1981.
[6] Bond, A. P. and Dundas, H. J. *Materials Performance*, **23**, 39–43, July 1984.

2

Cast high alloy stainless steels

R. F. Atkinson
Lake & Elliot Paramount Limited

The 1980s have been the most challenging and exciting years in the development and use of high alloy stainless steel for critical seawater applications. Many new alloys have appeared on the scene and there has been a revised interest in many of the older alloys which although not seawater 'proof' in all respects have, nevertheless, given good service over many years.

The bulk of the challenge came from North Sea oil developments which required castings of both extremely high quality and corrosion resistance. The very severe laboratory tests now available for determining pitting corrosion resistance, crevice corrosion resistance and stress corrosion resistance have shown the limitations of many of the common high alloy stainless steels for the very critical applications and this has resulted in the development of many new alloy types. Some of these new alloy types are not particularly well suited to the production of castings and indeed even wrought materials and production of satisfactory castings has involved considerable process development.

Systematic development, however, takes time and that is something oil companies rarely have available once a project has been given the go-ahead. In many cases, therefore, development and production have been concurrent, and whilst this can be very expensive it has also been advantageous in shortening the normal development cycles from years to months.

In this chapter it is proposed to review, briefly, the role and development of cast high alloy stainless steels for critical seawater applications, including some nickel base alloys which are finding increasing use in the very critical applications. Sand, investment and centrifugal castings will be considered and some of the interesting recent applications will be highlighted.

MARTENSITIC STAINLESS STEELS

The two most common alloys in use are alloys 1 and 2 in Table 2.1. Paralloy MM1 (CA6NM) was first promoted for use in seawater by George Fischer Limited over 20

Table 2.1 — Martensitic and precipitation hardening cast corrosion-resisting steels

Grade	Nominal composition									Minimum properties		
	C	Mn	Si	P	S	Cr	Ni	Mo	Cu	UTS MPa	0.2 YP MPa	Elong (%)
ASTM A743 CA-15 Paralloy M15	0.15	0.6	0.6	0.02	0.02	13	—	—	—	620	450	18
ASTM A743-CA-6NM Paralloy MM1	0.06	0.6	0.6	0.02	0.02	13	4	0.5	—	755	550	15
Paralloy MPH	0.05	0.5	0.6	0.02	0.02	17	4	—	2.5	1000	770	12
Paralloy MPH5	0.05	0.6	0.6	0.02	0.02	14	6	2.0	2.0	950	800	12

years ago [1] as an improvement over CA-15 in corrosion resistance, castability and weldability. It was used for 8-tonne cooling water impellers handling seawater for the Fawley power station and it was chosen for some of the well-head equipment of the more aggressive North Sea wells long before the NACE requirements on this material really came into effect. Discussion of the NACE requirements and the work done in production materials to meet the 23 Rockwell C hardness maximum has been the subject of several papers [2,3]. One of the major problems relating to the NACE requirement is the conversion of 23 Rockwell to Brinell, the standard conversion tables being for carbon or low alloy steels and, therefore, not accurate for CA6NM. The Rockwell C test is difficult on anything except square blocks, the preferred test for complex shapes is Brinell and the conversion factor is, therefore, critical. It is, however, normally just possible to meet the NACE requirements using the conventional conversion tables provided there is very tight control of carbon and silicon levels and a harden and double temper is carried out.

CA6NM 4-inch centrispun tube was recently supplied for sub-sea manifolds and bends of radii 3D to 5D were also produced from this tube without difficulty. All tubes supplied met all the NACE requirements and were 100 per cent ultrasonically tested, a factor which can influence the selection of this material for seawater service in preference to the austenitic alloys. This alloy can suffer pitting and crevice corrosion in seawater.

PRECIPITATION HARDENING STAINLESS STEELS

Alloys 3 and 4 in Table 2.1 are in regular use. Their use has not increased significantly in the past year for seawater applications. The only recent application is yacht winches and shackles. In this application the material is highly polished and seems to resist pitting corrosion, etc., very well. These alloys still find application in seawater service for pump and valve trim, where their high strength/hardness is of benefit.

AUSTENITIC STAINLESS STEELS

The two grades most commonly used are CF8M and CF3M (Table 2.2). They have been used extensively for pumps and valves handling seawater in spite of their poor

Table 2.2 — Austenitic cast corrosion-resisting steels

Grade		Nominal composition							Minimum properties		
	C	Cr	Ni	Mo	Cu	N		UTS	0.2 % proof MPa	Elong (%)	
ASTM A743 CF-8M Paralloy 3C	0.08	18	9	3				485	205	30	
ASTM A743 CF-3M Paralloy 3KLC	0.03	18	9	3				485	205	30	

resistance to crevice corrosion compared with the higher alloy materials. In some applications for castings crevice corrosion is not the major problem particularly where components are of a heavy section and in many cases crevices can be designed out. This type of alloy does pit in seawater under static conditions particularly where surfaces are shielded by marine growth. The emphasis on this material has been on improving the corrosion resistance as much as possible by reducing carbon content and applying better treatments. The effects of these two factors are dealt with in a paper by B. Warren [4]. Marginal improvements in the corrosion resistance of this type of stainless steel can be made by higher solution treatment temperatures and water quenching but this can cause problems of thermal distortion on the more intricate castings and the marginal improvements in corrosion resistance must be weighed against this. In spite of their limitations when exposed to the onerous corrosion tests of modern laboratories CF3M and CF8M will continue to be specified for many seawater applications.

DUPLEX AND 'SUPER' DUPLEX STEELS

There are many types of duplex stainless steels but the most common are listed in Table 2.3. The high strength of these materials compared with the austenitic grades CF8M and CF3M should be noted.

The pros and cons of the different grades have been discussed in many papers [5]. The author's company initially favoured the Paralloy 3FL composition which it first manufactured for the load bearing tubular lattice of Bush Lane House [6]. This grade is the equivalent of the most specified wrought grade of duplex stainless steel and has similar corrosion resistance generally to CF8M but superior resistance to stress corrosion and much higher strength.

The higher chromium grades of Duplex such as Ferralium have much better resistance to pitting and crevice corrosion in seawater but may require post weld heat

Table 2.3 — Duplex and super duplex corrosion-resisting steels

Alloy	Nominal composition							Min. props.		
	C	Ni	Cr	Mo	W	Cu	N	UTS MPa	0.2%PS MPa	El %
Ferralium 3SC	0.06	6.0	25.0	3.0	—	2.5	0.17	725	450	20
Ferralium 255	0.06	5.0	25.0	2.5	—	2.5	0.17	725	485	20
Zeron 25	0.03	6.5	25.0	2.5	—	0.5	0.18	685	480	29
Zeron 100	0.03	7.0	25.0	3.5	0.8	0.75	0.25	750	500	34
Paralloy 3FLC	0.03	5.0	22.0	3.0	—	—	0.14	680	470	35
Fermanel	0.06	8.5	27.0	3.0	—	1.0	0.22	750	450	25
FMN	0.06	5.0	25.5	2.0	—	—	0.20	695	495	36

treatment if fabricated. With Paralloy 3FL, however, satisfactory weld properties have been obtained on welding 38 mm centrispun tube and castings without post weld heat treatment. Four-inch, 6-inch and 10-inch schedule 60 centrispun pipe has been supplied for the fabrication of sub-sea manifolds.

Zeron 100 and Fermanel are what is now known as 'Super' duplex stainless steels and have even better corrosion resistance in seawater but there is little data available on the performance of these materials in service, as they are relatively new. The compositional differences between the 'Super' duplex and duplex grades containing 25 per cent chromium are small and it is felt that the differences in performance will be small.

It is anticipated that the duplex steels will gradually replace CF8M in seawater applications. There are, however, some engineers who are still not satisfied with the corrosion resistance of duplex welds and trials are still in progress with more highly alloyed welding consumables.

HIGH ALLOY AUSTENITIC STAINLESS STEEL

This is a class of material which has been growing in importance over the past few years. Typical compositions are shown in Table 2.4. The first of these alloys has been in the American standards for 20 or 30 years. It has proved difficult to manufacture in heavy sections, is prone to crack on welding and has been the subject of much research and development by the Steel Foundry Society in America [7,8]. In efforts to improve the castability/weldability of this alloy investigations have been carried out into the effects of impurity but no positive way has been found of improving the alloy. Nevertheless, satisfactory castings can be made in this alloy provided care is taken in foundry methoding to ensure that weld repair is kept to an absolute minimum. A better alloy is the second one shown in Table 2.4 which is pioneered by INCO as Alloy 862 [9]. It has better castablility and weldability than CN7M and has

Table 2.4 — Austentitic high alloy cast corrosion-resisting alloys

Grade	Nominal composition							Minimum properties		
	C	Cr	Ni	Mo	Cu	Nb	N_2	UTS MPa	0.2% Proof MPa	Elong %
ASTM A743 CN7M Paralloy P20	0.05	20	29	3	4	—	—	425	170	35
Paralloy 6NLC (IN 862)		21	25	5	—	—	—	425	170	30
Paralloy Avesta 254 SMO	0.02	20	18	6	0.7	—	0.2	500	250	35
Paralloy P40	0.03	21	42	3	2	0.9	—	480	205 (static) 240 (centrispun tube)	25

been extensively used for pumps and valves handling both seawater and waters containing much higher percentages of chloride.

The need to develop materials with better crevice pitting corrosion resistance has resulted in the emergence of several alloys containing much higher percentages of molybdenum. Typical of these is Avesta 254SMO, the third alloy, which also contains nitrogen thus improving pitting corrosion resistance and also strength. Strength can be an important factor in austenitic stainless steels as the majority of them are relatively weak when compared with duplex stainless steels. Cast 254SMO can be readily produced in the foundry and it is best welded with high nickel electrodes, the resulting welds having good corrosion resistance even in the as-welded condition. This alloy has great potential for critical seawater applications in the future [10].

Paralloy P40 is the cast version of IN825 developed by the International Nickel Company. The cast version contains niobium instead of titanium and has slightly lower mechanical properties than the wrought version. In both centrispun tube and sand castings, however, the minimum proof strength specified in ASTM B424 can be met. Careful control of all impurities is necessary in this material to obtain good ductilities and 180° bends can be obtained from 30 mm diameter round bars in both the as cast and heat-treated condition. Satisfactory 90° bends have been produced in this material from 6-inch centrispun tube. Although there is considerable interest in the cast form of this material at present, it is expected that materials such as 254SMO and duplex steels will supersede it.

Fig. 2.1 — Nickel-based Paralloy 625 castings for Shell Fulmar Field.

CAST NICKEL BASED ALLOYS

The alloy in which there has been particular interest is Cast 625. The compositions of the two grades produced here are shown in Table 2.4, the compositions being essentially the same as that of IN625 developed by the International Nickel Company. This alloy is not only exceptional in terms of corrosion resistance but also has good high temperature properties. It was for the latter purpose that the author's Company first manufactured large quantities of tube and small sand castings in high temperature grade Paralloy HTN625. Recently, large quantities of sand castings and centrispun tube up to 1200 kg in weight have been produced in the low carbon version CRN 625 for North Sea oil applications.

Recent tests on CRN 625 have shown that it has the same excellent corrosion

Fig. 2.2 — Manifolds for Marathon Brae 'B' platform fabricated from Paralloy P40(825) centrispun tube. Courtesy of Scomark Ltd.

resistance as the wrought product [11] and more recent unpublished work has tended to show that in this alloy carbon content has little bearing on corrosion resistance, the high carbon version being as good as the low carbon version.

The alloy is not easy to cast and considerable difficulties were experienced in the initial stages requiring development of several special techniques within the foundry. Refinement of these techniques has lead to a steady improvement in the quality of castings and recently it has been possible to produce nuclear quality valve castings within 14 working days, something which would have been quite impossible in the early stages of development.

The alloy is complex metallurgically and the mechanical properties are sensitive to impurity levels. CRN 625 is supplied in the solution-treated condition and in this form has lower mechanical properties than the wrought hot rolled and annealed form of the alloy. CRN 625 can be readily welded in heavy sections and the welds will withstand 180° side bends.

No section on the nickel-based alloys would be complete without mentioning Hastelloy (Cabot Corporation Register Trade Mark). There has always been a steady demand for Hastelloy castings, particularly Hastelloy Ccastings, and Cabot Corporation have done development work to produce an alloy which has satisfactory properties. The alloy developed is Hastelloy C4C, the composition of which is shown in Table 2.5. The new alloy has a very low carbon content and contains no tungsten

Table 2.5 — Nickel base corrosion-resisting alloys

| Grade | C | Cr | Ni | Mo | Nb | Fe | Min. prop. | | |
							UTS MPa	0.2% proof MPa	Elong (%)
Paralloy CRN 625	0.03	21	BAL	8.5	3.5	5.0	500	250	25
Paralloy HTN 625	0.10	21	BAL	8.5	3.5	5.0	520	270	20
Hastelloy C-4C[a]	0.02	16	BAL	16		2.0	495	275	35
ASTM A494 C12 MW2	0.07	18	BAL	18		3.0	495	275	25

[a] Cabot Corp. Registered Trade Mark.

and because of this has very high ductility with elongations of the order of 40 per cent after solution treatment. This alloy also has excellent resistance to crevice corrosion [11] and is similar to ASTM Grade C12MW2, although in Hastelloy C4C impurities are more closely controlled, giving better corrosion resistance and ductility.

CONCLUSION

There has been in the past few years a big improvement in the range of cast high alloy steels available and comparable alloys can be produced for virtually all wrought specifications. The quality of castings now produced has also improved significantly due to tighter control of impurity levels, improved semi-automatic moulding processes and the introduction of the new ladle and pouring techniques developed by the Steel Castings Research and Trade Association.

Centrispun tube has been used for over 20 years for high temperature oil and chemical processing plants and there are at last signs that this high integrity product is being used for more of the critical seawater applications. Because of the centrifugal action used in the production of such tube it is extremely clean and is, therefore, ideal for further forming operations such as bending. Dimensional tolerances are better than those obtained from either rolled and seam welded tube or extruded tube and it contains, of course, no seam weld. Engineers and designers are now using more of this product for corrosion-resisting applications in the North Sea.

REFERENCES

[1] Gerber, G. E. and Trautwein, A. CA6NM. New developments based on 20 years experience, *Steel Foundry Facts*, No. 343, February 1981–2.

[2] Nalbone, C. J. Effect of carbon content and tempering treatment on mechanical properties and sulphide stress-corrosion cracking resistance of AOD-refined CA6NM, *ASTM Symposium* 1980. Miami. STP 756.

[3] Fischer, R. B. and Larsen, J. CA6NM to RC22 maximum *Steel Foundry Facts*, No. 343, February 1981–2.

[4] *Effect of Metallurgical Quality on Corrosion Resistance of Cast Stainless and Nickel Base Alloy*, The International Nickel Co. Inc.

[5] Duplex stainless steel ASM conference, St Louis, October 1982.

[6] Unique building application: Welding Institute Conference, September 1982.

[7] Larsen, J. A. New developments in high alloy cast steels, Steel Castings Research and Trade Association, Research Conference 1984

[8] Cieslar, M. J. and Savage, W. F. *Hot cracking of CN7M*, Steel Founders of America. T&O Conference 1984.

[9] Forbes, R. M. *A Cast Stainless Steel for Chloride Environments*, American Foundry Society Conference, Montreal 1973.

[10] Wallen, B. *Seawater Resistance of a High Molybdenum Stainless Steel*, NACE Corrosion Conference, January 1977.

[11] Hack, H. P. Crevice corrosion behaviour of molybdenum containing stainless steel in seawater, NACE Corrosion '82 Conference, Paper 65.

3

Welding high alloy stainless steels

T. G. Gooch
The Welding Institute, Abington Hall, Abington, Cambridge CB1 6AL

INTRODUCTION

At the present time, recently developed high alloy stainless steels specified for service under seawater conditions [1–3] can be divided into three types, viz.:

1. high alloy fully austenitic,
2. duplex ferritic/austenitic, and
3. fully ferritic extra low interstitial (ELI) ferritic grades.

This chapter considers the salient fusion welding characteristics of materials in each alloy class with the aim of giving general guidelines, and, in particular, identifying potential problem areas.

AUSTENITIC STEELS

In principle, any arc process can be employed but most practical experience has been obtained with the manual metal arc (MMA) and tungsten inert gas (TIG) processes. Essentially, conventional austenitic stainless steel welding techniques are appropriate, and, from the point of view of making a weld, the main recommendation [4] is employment of minimal heat input levels commensurate with obtaining good fusion and penetration. The aim is to minimize the occurrence of solidification or liquation cracking and any loss of corrosion resistance from carbide or intermetallic precipitation in the weld metal or surrounding heat affected zone (HAZ).

Solidification cracking is widely associated with fully austenitic weld metals as are inherent in the various high alloy compositions marketed for seawater applications. However, accepting that these materials have largely been welded with careful attention to cleanliness and welding procedure, this form of cracking does not seem to have been a major problem, possibly because of a beneficial effect of the high steel molybdenum contents. None the less, caution is necessary since crater cracking may occur, and attention must be paid to arc extinction procedure. Because castings tend to have a coarse grain size, they may be sensitive to grain boundary liquation cracking in the HAZ adjacent to the fusion boundary. Both grain size and impurity element content should be minimized to reduce the problem.

Intercrystalline attack is, in fact, unlikely in view of the low carbon contents of commercial alloys, although, in fairly heavy multi-pass welds, intermetallic formation may have an adverse effect on corrosion resistance if prolonged thermal cycles are involved. In this regard, maintenance of low interpass temperatures is essential, preferably below 100°C for the high molybdenum alloys.

The main difficulty with high alloy austenitic grades has been the achievement of weld metal corrosion resistance fully matching that of the parent steel (Table 3.1) [5].

Table 3.1 — Results of crevice corrosion tests [5] on high alloy austenitic stainless steel butt welds in slowly flowing seawater at 70°C

Parent steel	Welding process	Consumable type	Depth of attack		Number of attacked samples
			Weld metal	Parent material	
316	MMA	316	0.05	0.08	3 out of 3
20Cr/18Ni/ 6Mo/Cu/N	TIG	Autogenous[a]	<0.1	0	1 out of 3
	MMA	60Ni/20Cr/9Mo	0	0	0 out of 3
	MIG	60Ni/20Cr/9Mo	0	0	0 out of 3

[a] Postweld heat treated at 1150°C for 5 min.

Inevitably, during solidification segregation of chromium and molybdenum takes place from the solid to liquid phase, resulting in regions of weld metal that are appreciably depleted in these elements, and hence causing local reduction in corrosion resistance. The effect is not always of practical import, but segregation can be quite marked and caution must attend the use of autogenous welding or of matching composition consumables (Fig. 3.1) [6].

To overcome this limitation, it has been advised that welds be postweld heat treated at *c.* 1150°C, but even at this temperature diffusion of chromium and molybdenum is fairly slow and complete homogenization does not occur. Thus the most common recommendation is that, as far as possible, welding should be carried out with a non-matching nickel-based consumable having a high molybdenum content. Although full parent steel corrosion resistance is not achieved, it is doubtful whether the difference is of great significance. When using such consumables, excessive dilution by parent material must be avoided: it is essential that sufficient

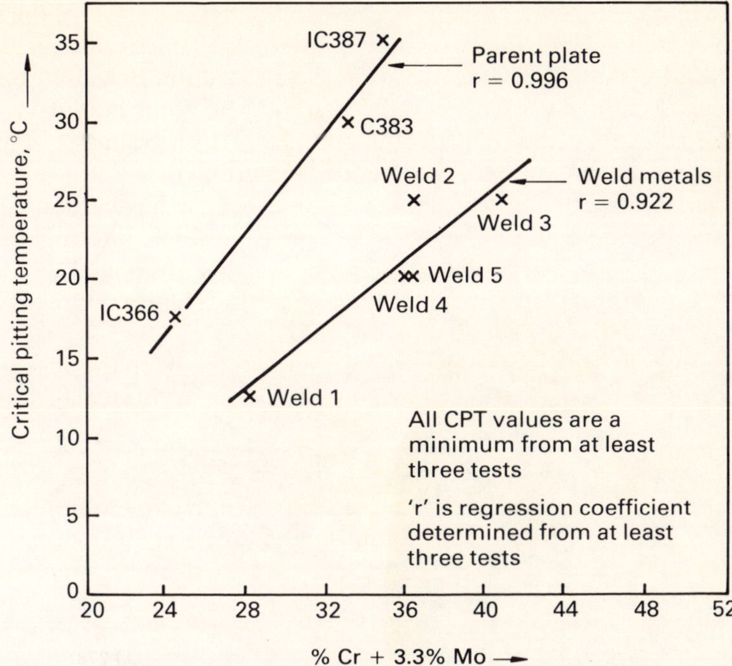

Fig. 3.1 — Critical pitting temperature in 10%FeCl$_3$ versus %Cr+3.3%Mo for high alloy austenitic steels and MMA weld metals [6].

filler be added and joint preparations should be designed to ensure that this is achieved. As with all stainless steels, careful postweld cleaning to remove residual scale is advisable to obtain maximum corrosion resistance.

FERRITIC/AUSTENITIC STEELS

In parent material form, these alloys contain a duplex structure with approximately 50/50 ferrite/austenite balance. However, on welding, transformation of ferrite takes place at high temperatures with associated grain growth, and this can significantly reduce the local toughness and corrosion resistance. Current alloys are designed to give reformation of austenite in the weld area on cooling (Fig. 3.2) [7], and in large part commercial materials show good weld area properties despite the major phase charges that take place during welding.

Most arc welding processes have been successfully applied to ferritic/austenitic steels. Even with current alloy design, it is necessary that reasonably high input levels are employed during welding. If this is not the case, the rapid cooling from peak temperatures suppresses austenite formation and can lead to substantial loss in properties (Fig. 3.3). At the same time, excessive arc energy should be avoided, since protracted cooling encourages precipitation of carbonitride and intermetallic phases, with associated loss in corrosion resistance. In general, low interpass

Fig. 3.2 — HAZ transformation in duplex stainless steels [7]: (a) 26%Cr/5%Ni/1.5%Mo (older steel type, ×100; (b) 18%Cr/5%Ni/2.5%Mo/N (newer steel type), ×500.

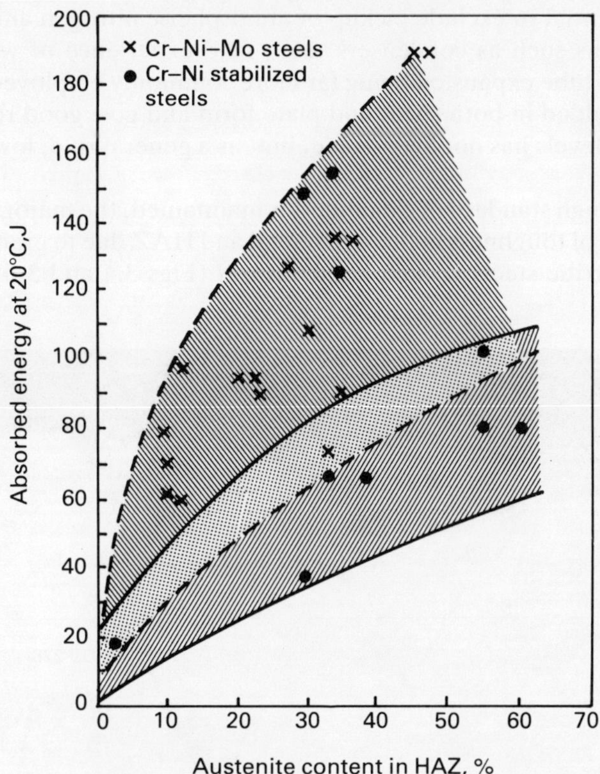

Fig. 3.3 — General relationship HAZ austenite content and Charpy impact energy at +20°C for duplex stainless steels [7].

temperatures are preferred, certainly below 300°C, to minimize the time spent in the '475°C embrittlement' temperature range.

Final austenite levels tend to be lower in matching composition weld metal than in the HAZ and for this reason it is common practice to employ consumables with slightly higher nickel content than the base steel to encourage austenite development in the weld metal. Weld area austenite levels should be determined during procedural testing at, say, 1050°C. They can be applied to restore a 50/50 ferrite/austenite ratio, but this is not always practicable, and, it must be emphasized, is seldom necessary given appropriate attention to material composition and welding procedure.

FERRITIC STEELS

In terms of making a weld, there is no major problem with fully ferritic grades but to guarantee good weldment toughness and corrosion resistance extraneous contamination must be minimized [1]. It is therefore mandatory to pay stringent attention to cleanliness: welding virtually always employs an inert gas-shielded process and good gas shielding is essential to exclude pickup of atmospheric nitrogen and oxygen. In terms of applications such as condensers, etc., field experience of welding these materials is limited, tube expansion being far more commonly employed. However, the steels can be welded in both sheet and plate form and give good results. Close study of heat input levels has not been made, but, as a general rule, low heat inputs are to be preferred.

Provided that a high standard of cleanliness is maintained, the major drawback to welding is some loss of toughness in the weld metal and HAZ due to grain coarsening inherent in fully ferritic steels at high temperatures (Figs 3.4 and 3.5) [8]. This is

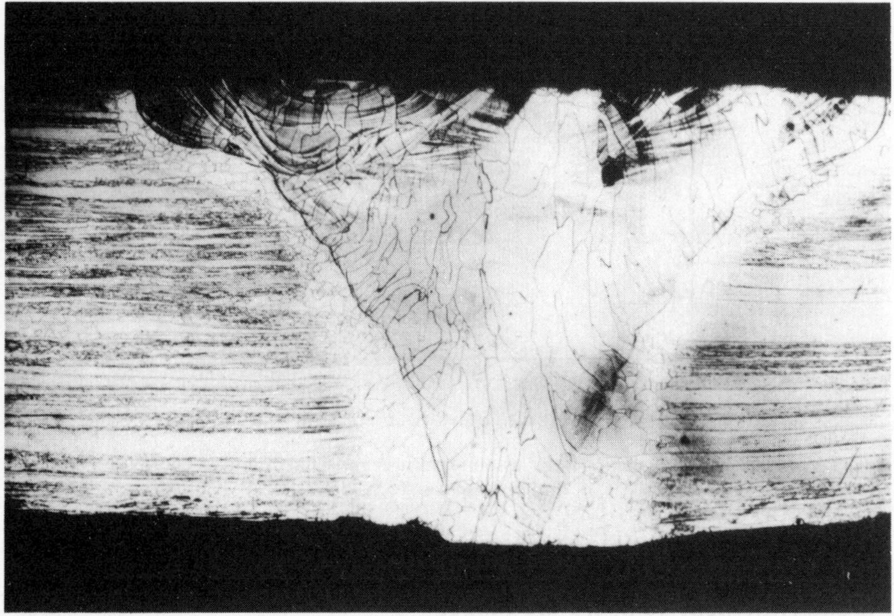

Fig. 3.4 — TIG weld in 13 mm 26%Cr/1%Mo steel ELI ferritic steel made with matching filler wire, ×5.

normally more significant with plate steel than with sheet. The effect may not be important for many applications, but should be recognized during design and operation of plant.

Both older and lower alloy grades (such as 26Cr/1Mo and 18Cr/2Mo respectively) have been welded with austenitic consumables, largely for improved weld metal toughness, but current steels with higher chromium and/or molybdenum levels are normally welded autogenously or with matching filler to obtain weld metal corrosion resistance similar to that of the parent material. A range of consumable compositions can be employed for dissimilar metal joints to other types of steel, high molybdenum nickel-base filler probably being most often preferred.

It can be remarked that segregation of chromium and molybdenum in weld metal

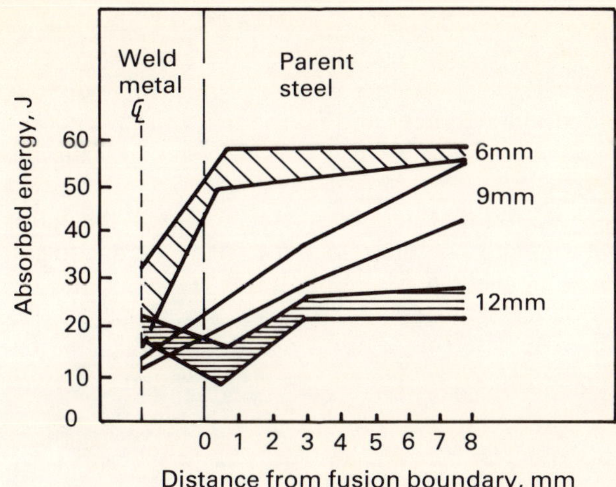

Fig. 3.5 — Room temperature Charpy impact toughness of TIG welds in 28%Cr/2%Mo/4%Ni/ Nb steel [8]. (Note that data for 6 and 9 mm plate are from subsize samples.)

is very much less in these materials, with primary solidification to ferrite, than in the high alloy austenitic grades. In consequence, weld metal coring does not pose a particular problem, and corrosion resistance of weld metals can be very close to that of parent steel.

CONCLUDING REMARKS

There is no doubt that satisfactory welds can be reliably made in steels in the three alloy classes considered above. At the same time, in all cases welding can have marked adverse effects and this must be recognized during design and installation of any plant built in high alloy materials for seawater service. To some degree, there is a common link in that it is often likely that the weld metal will be limiting on joint performance, and there is a strong case wherever possible to use automated welding procedures to minimize operator variations and ensure compliance with the above guidelines.

REFERENCES

[1] *Proceedings of Symposium: Advanced Stainless Steels for Seawater Applications*, Piacenza, Italy, 1980, Climax Molybdenum Co.
[2] *Proceedings of Conference: Duplex Stainless Steels*, St Louis, USA, 1982, ASM.
[3] *Proceedings of Conference: Stainless Steel 77*, London, UK, 1977, Climax Molydenum Co.
[4] Bernhardsson, S. O., Mellstrom, R. and Tynell, M. (in Ref. 1).
[5] Wallen, B. (in Ref. 1).
[6] Hale, G. E. (unpublished work).
[7] Gooch, T. G. (In Ref. 2).
[8] Oppenheim, R. (in Ref. 3).

4

Galvanic compatibility of selected high alloy stainless steels in seawater

E. B. Shone and P. Gallagher
Shell Research Ltd., Thornton Research Centre, P.O. Box 1, Chester CH1 3SH

INTRODUCTION

The attraction of using high alloy stainless steels for seawater handling systems on offshore platforms has been outlined in another paper [1].

Seawater handling systems are made up of pipework, valves, heat exchangers, pumps and other smaller components. During construction, commissioning and service the seawater within such a system may range from stagnant to fast flowing and crevices will exist. The temperature of the seawater can be expected to vary between 5°C and 30°C and after passing through a heat exchanger this could, under most instances, be increased by up to 10°C. However, temperature increases of up to 25°C are not unusual. For any seawater system to operate successfully and require minimal maintenance, we believe that it must be designed as a system from compatible components and not built up from collections of what may well be well-engineered components that are incompatible from a corrosion viewpoint.

It is inevitable that within a seawater system more than one alloy type will be chosen because of varying strength requirements or even availability in a particular form. Electrical isolation of sections of the system cannot be relied upon offshore since electrical safety requirements (earthing) predominate on a steel structure. It is therefore essential that materials selected for the seawater systems are galvanically compatible. With this in mind we began a research programme aimed at establishing the galvanic compatibility of various materials. However, before the compatibility of the materials could be meaningfully established we believed that it was first necessary to determine the intrinsic corrosion properties of the alloys.

EXPERIMENTAL

A test programme was carried out on a range of alloys that could be considered for possible use in seawater handling systems. Two types of screening test were initially carried out in natural seawater, these were:

(a) Crevice corrosion tests using a multiple crevice washer test.
(b) Potentiodynamic tests.

In the above tests 35 different alloys were evaluated. As a result of these two testing procedures the alloys were ranked in order of their susceptibility to crevice corrosion when exposed to seawater as single alloy specimens. A selection of the results is summarized in Chapter 8 (Fig. 8.3) in the form of a histogram of the degree of corrosion occurring in terms of crevice depth and area attacked over a period of 100 days. (For clarity these results are repeated as Fig. 4.1 in this chapter).

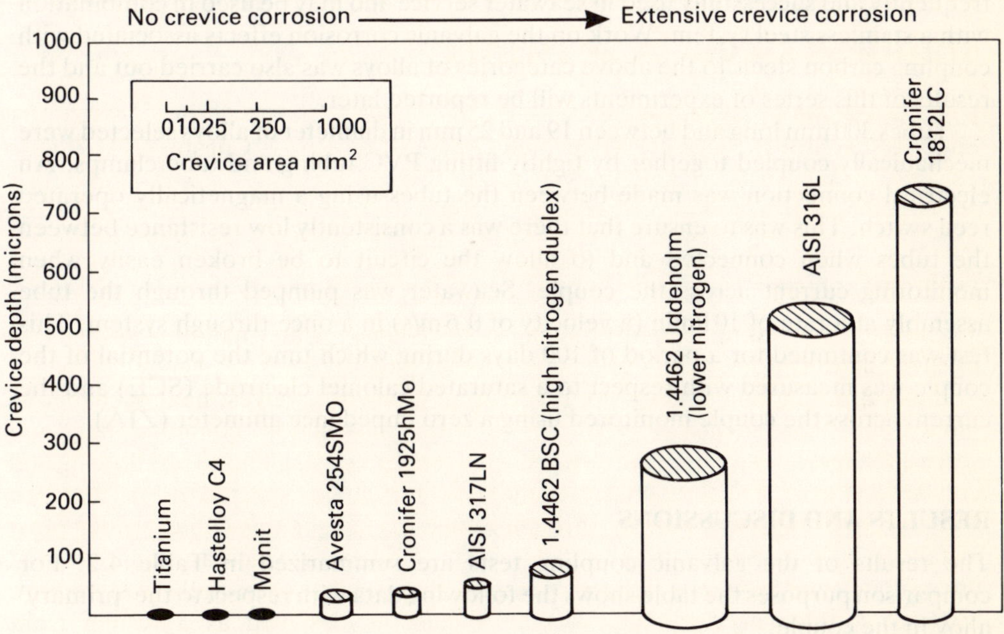

Fig. 4.1 — Crevice depth and area of attack on alloys after 100 days exposure.

Following this work a series of tests was conducted in which alloys were exposed to natural flowing seawater whilst electrically coupled. The aim of this series of experiments was to determine which pairs of alloys were compatible and could be used together in a total system without causing galvanically stimulated corrosion to occur on either alloy in the couple. To examine the effect of galvanically coupling all combinations of these alloys would have required a large number of tests and this would have been impracticable. Thus to reduce this number to a realistic level the alloys were grouped together into the basic categories listed below:

1. Titanium.
2. Superferritic stainless steel.
3. Nickel-based alloys.
4. High alloy austenitic stainless steel.
5. Duplex stainless steel.
6. 'Conventional' austenitic stainless steel.
7. 90/10 copper/nickel.

Samples representative of each category were tested in combination with each other. The trade names of materials selected as being representative of the various categories and their analyses are given in Table 4.1. The 90/10 copper/nickel alloy was included because this is the most frequently used alloy in seawater systems on offshore platforms, so it was decided to ascertain the compatibility of this alloy with the stainless steels since it is probable that at some stage these materials could be coupled together. In a similar way titanium was included because this alloy is frequently and successfully used in seawater service and may be used in combination with a stainless steel system. Work on the galvanic corrosion effects associated with coupling carbon steels to the above categories of alloys was also carried out and the results of this series of experiments will be reported later.

Tubes 300 mm long and between 19 and 25 mm in diameter of alloys selected were mechanically coupled together by tightly fitting PVC tubing and tube clamps. An electrical connection was made between the tubes using a magnetically operated reed switch. This was to ensure that there was a consistently low resistance between the tubes when connected and to allow the circuit to be broken easily when monitoring current across the couple. Seawater was pumped through the tube assembly at a rate of 10 l/min (a velocity of 0.6 m/s) in a once-through system. This test was continued for a period of 100 days during which time the potential of the couple was measured with respect to a saturated calomel electrode (SCE) and the current across the couple monitored using a zero impedance ammeter (ZIA).

RESULTS AND DISCUSSIONS

The results of the galvanic coupling tests are summarized in Table 4.2. For comparison purposes the table shows the following data with respect to the 'primary' alloy in the couple:

 (i)   maximum depth of corrosion;
 (ii) total depth of corrosion;
(iii) intrinsic crevice depth of uncoupled samples;
(iv) the maximum galvanic corrosion current measured during exposure.

From Table 4.2 it can be seen that none of the three alloy groups, titanium, superferritic stainless steels or high molybdenum stainless steels, is significantly affected by coupling any of the other alloys to them. In the tests with the nickel-based alloy Nicrofer 6020hMo as the 'primary' alloy, the Nicrofer 6020hMo tubes showed a surface etching effect as they did in the uncoupled tests where this effect was

Table 4.1 — Trade names and chemical analyses of alloys from the various categories

Alloy trade name	Category	Chemical composition (%w)										
		Cr	Ni	Mo	N	Mn	C	Si	P	S	Cu	Others
245 SMO[a]	High alloy austenitic	19.3	18.3	5.95	0.215	0.50	0.017	0.56	0.025	0.003	0.70	Nb-0.04
Cronifer 1925hMo[b]	High alloy austenitic	20.65	24.75	6.24	0.134	1.43	0.014	0.29	0.022	0.003	0.80	—
Monit[c]	Superferritic	25.5	4.0	4.02	0.023	0.25	0.014	0.29	0.02	0.003	—	—
Cronifer 1812LC[b]	'Conventional' austenitic 316L	17.2	12.8	2.8	—	1.77	0.016	0.30	0.029	0.009	—	—
Cronifer 2205LCN[b]	Duplex	22.3	5.5	2.77	0.118	1.73	0.01	0.31	0.023	0.006	—	—
Nicrofer 6020hMo[b]	Nickel based	21.85	64.1	8.87	—	0.01	0.029	0.08	0.003	0.005	—	Ti-0.2 Nb-3.4 Co-0.06
Kunifer 10[d]	Copper nickel	—	10	—	—	0.8	—	—	—	—	Bal	Fe-1.6
		O_2	N_2	C	Fe	H_2						
Titanium IMI125[d]	Titanium	0.11	<0.11	<0.01	0.06	45 ppm						

Manufacturers: [a]Avesta, [b]V.D.M., [c]Nyby Uddeholm, [d]I.M.I.

Table 4.2 — Galvanic coupling tests

Alloy category	Coupled to	Max. depth of corrosion on primary alloy	Total area of corrosion on primary alloy, mm^2	Intrinsic (uncoupled) corrosion depth	Maximum galvanic current, μA
Titanium	Nicrofer 6020hMo	0	0	0	−150
	Monit	0	0	0	−4
	254 SMO	0	0	0	−8
	Cronifer 2205LCN	0	0	0	−12
	Cronifer 1812LC	0	0	0	−100
Superferritic	Titanium	<10	10	<10	+4
	Nicrofer 6020hMo	0	0	<10	−38
	254 SMO	0	0	<10	−2
	Cronifer 2205LCN	0	0	<10	−66
	Cronifer 1812LC	0	0	<10	−275
High molybdenum austenitic	Titanium	0	0	0	+8
	Nicrofer 6020hMo	0	0	<10	−35
	Monit	0	0	0	+2
	Cronifer 2205LCN	25	28	<10	−200
	Cronifer 1812LC	0	0	<10	−400
	90/10 Cu/Ni	0	0	<10	−1500
Nickel based	Titanium	10	785	25	+150
	Monit	10	250	25	+38
	Cronifer 1925hMo	25	465	25	+35
	Cronifer 2205LCN	20	700	25	−100
	Cronifer 1812LC	10	108	25	−95
Duplex 2205LCN	Titanium	250	115	200	+12
	Nicrofer 6020hMo	480	130	200	+100
	Monit	130	90	200	+66
	Cronifer 1925hMo	1000	75	200	+200
	Cronifer 1812LC	25	120	200	−500
	90/10 Cu/Ni	50	305	200	−1000
'Conventional' austenitic (AISI 316L type)	Titanium	850	95	500	+100
	Nicrofer 6020hMo	650	170	500	+95
	Monit	1000	908	500	+275
	Cronifer 1925hMo	500	165	500	+400
	Cronifer 2205LCN	850	305	500	+500
	90/10 Cu/Ni	900	65	500	−950

attributed to the loss of chromium and niobium in the surface layers during manufacture.

In the case of the duplex stainless steel, Cronifer 2205LCN, there was a noticeable increase in pit depth and area when coupled to titanium, Cronifer 1925hMo or Nicrofer 6020hMo. This increased crevice corrosion on the duplex stainless steel is presented in Fig. 4.2. The depth of crevice corrosion on the duplex

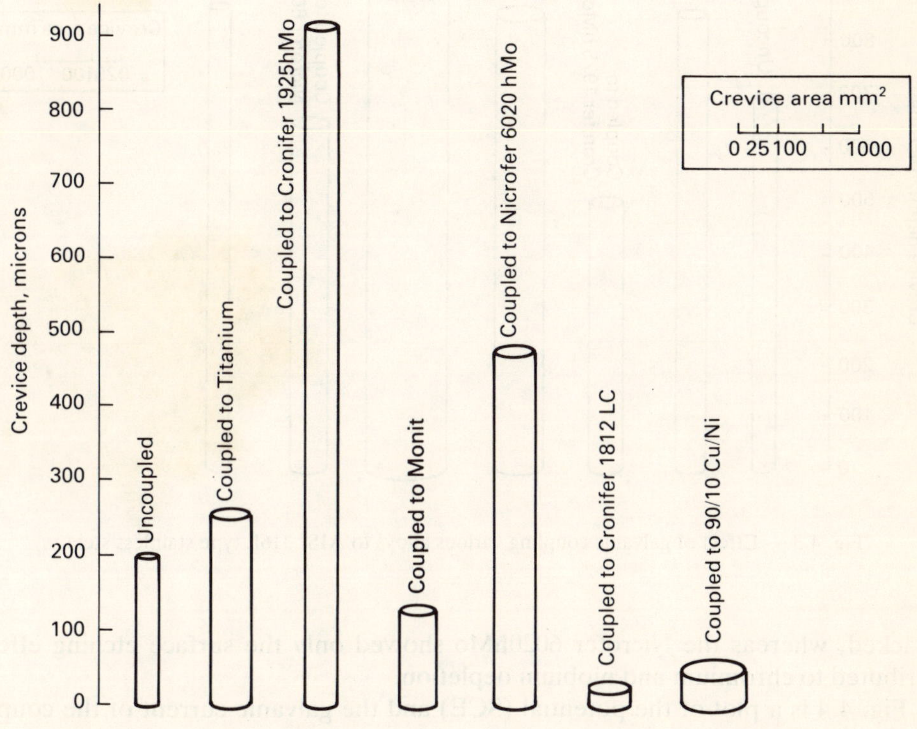

Fig. 4.2 — Effect of galvanic coupling various alloys to duplex stainless steel.

stainless steel was slightly reduced from its 'intrinsic' value when it was coupled to the superferritic alloy Monit. However, in this instance, the area was significantly higher, i.e. the total material loss was greater. When the duplex stainless steel was coupled to either the Cronifer 1812LC or the 90/10 copper/nickel material then the depth of crevice corrosion was significantly lower than for the uncoupled duplex. This is probably a result of these two alloys imparting a degree of cathodic protection to the duplex stainless steel. Near to the bimetallic junction the copper/nickel was noticeably more corroded.

The protective action of the copper/nickel was also apparent to some extent when coupled to the Cronifer 1812LC, since this was the one case when the area of corrosion on the Cronifer 1812LC was not increased. In all other cases there was a marked increase in the area of attack (Fig. 4.3). The Cronifer 1812LC was severely

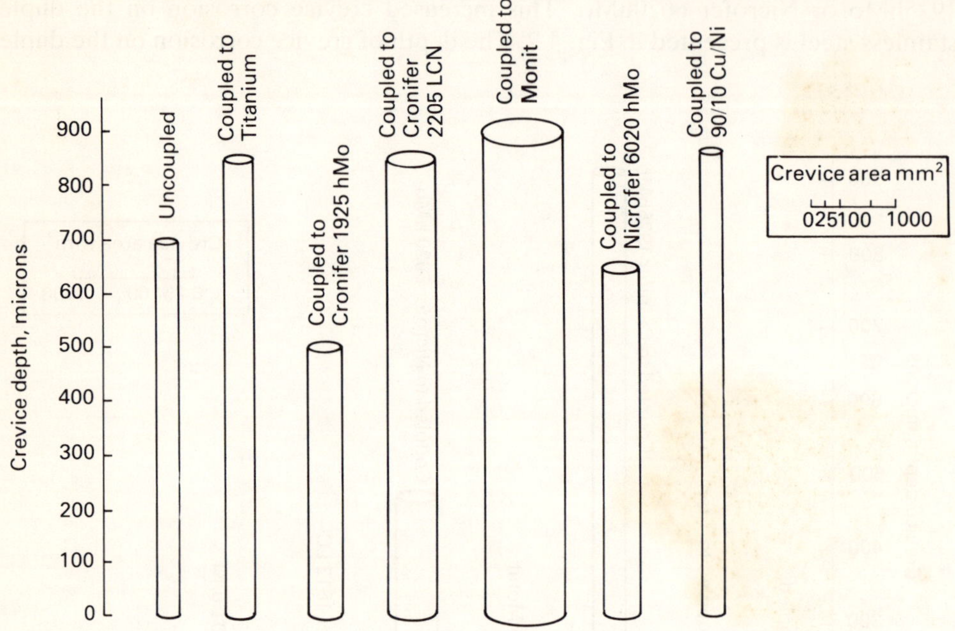

Fig. 4.3 — Effect of galvanic coupling various alloys to AISI 316L type stainless steel.

attacked, whereas the Nicrofer 6020hMo showed only the surface etching effect attributed to chromium and niobium depletion.

Fig. 4.4 is a plot of the potential (SCE) and the galvanic current of the couple between Cronifer 1912LC and titanium. Examination of the two plots shows that a drop in potential corresponds with an increase in current across the couple. If no significant corrosion occurs this rise in galvanic current is not apparent. An example of this is seen in Fig. 4.5 which shows the potential and current measured when 254SMO is coupled to Monit over a 100-day period. This was a consistent pattern which was observed in all cases. The fall in potential is attributed to the onset of crevice corrosion on one of the alloys in the couple.

From this work it can be concluded that:

1. The materials which are intrinsically not susceptible to crevice corrosion when exposed to seawater and not coupled to another alloy are also resistant to this form of corrosion when coupled even to more noble materials.
2. For alloys which are intrinsically susceptible to crevice corrosion the propagation rate of this form of attack can be increased by coupling noble alloys to them.

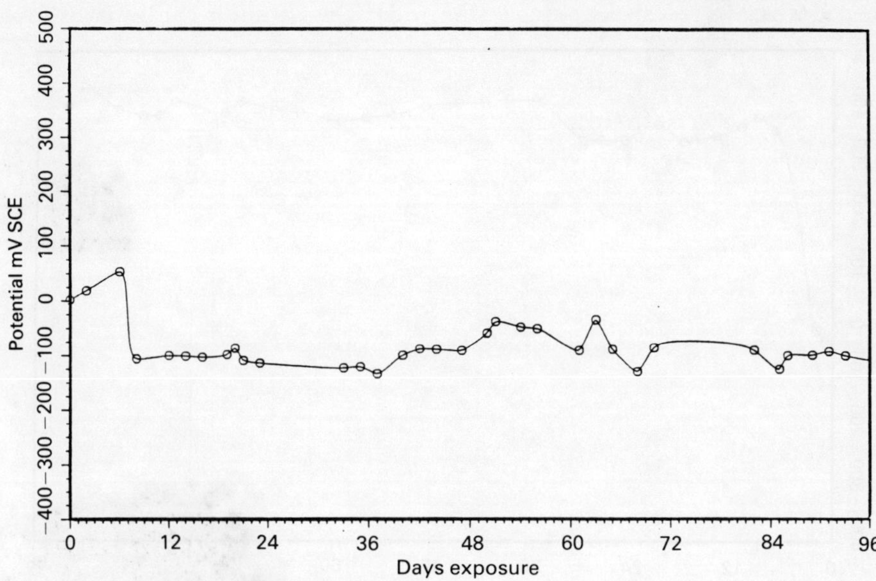

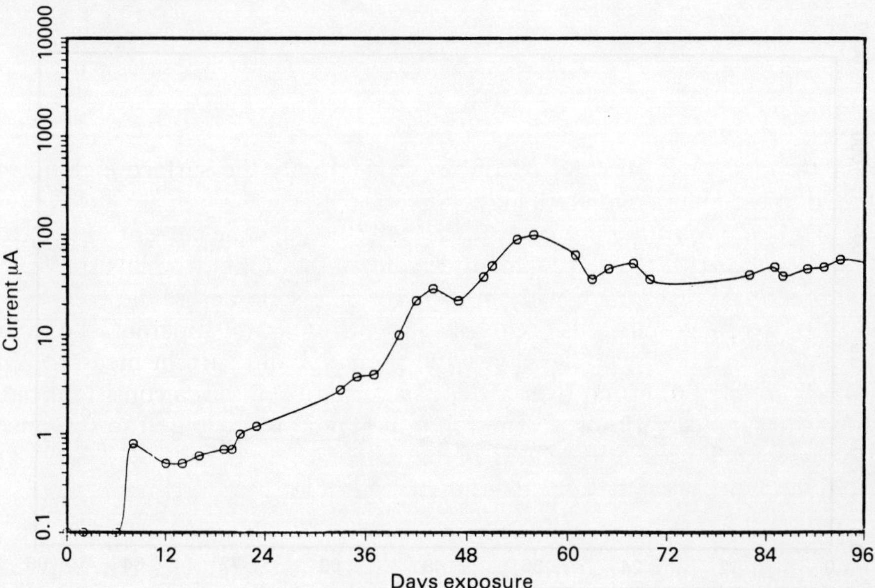

Fig. 4.4 — Titanium coupled to Cronifer 1812LC.

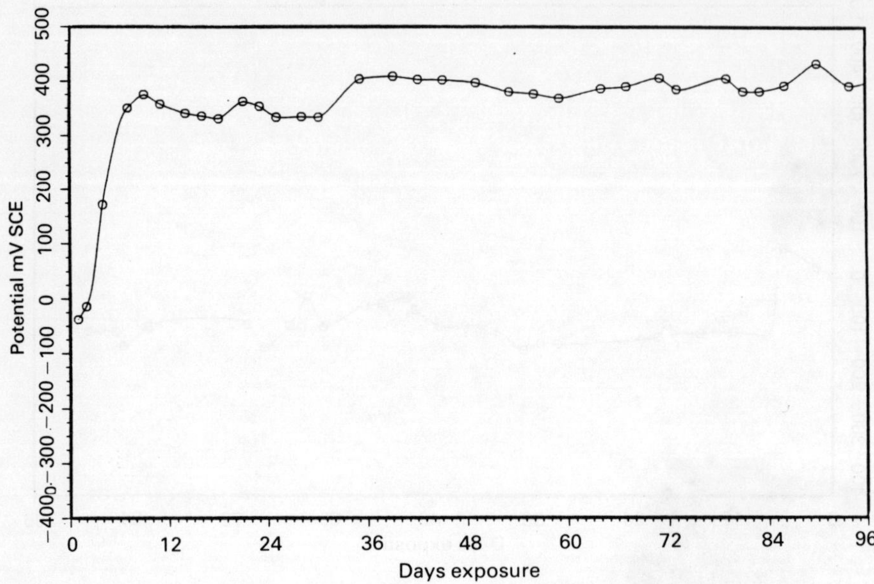

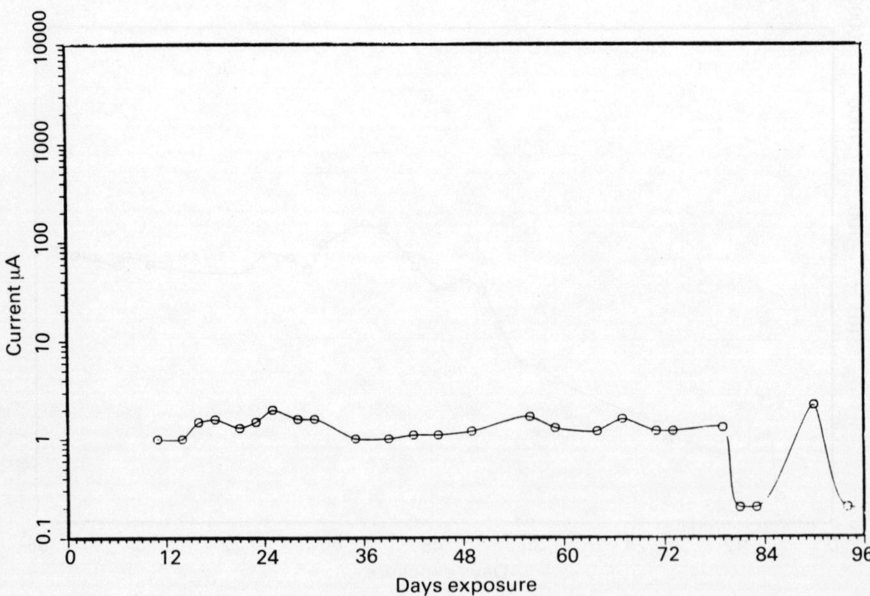

Fig. 4.5 — Monit coupled to 254SMO.

3. A correlation between potential drop and current increase across the coupled alloys is indicative of the onset of crevice corrosion.

REFERENCE

[1] Malpas, R. E., Shone, E. B. and Gallagher, P. Corrosion and Chlorination in Materials for Offshore Seawater Systems, Chapter 8, this volume.

Part II
Chlorination

5

Marine fouling control and chlorination†

J. W. Whitehouse
Central Electricity Research Laboratories, Leatherhead

Marine fouling is that community of organisms found growing on permanently or intermittently submerged surfaces of man-made objects in seawater. The growth of this community usually interferes with the efficient use of the surfaces, for example a reduction in ship's speed, reduction in internal diameter of cooling water pipes or the encrusting of support legs of oil rigs. The point about marine fouling is that it is universal and its control is a battle which has been fought as long as man has been crossing the seas or using seawater to enhance his life-style [1]. The purpose of this chapter is to introduce the subject of marine fouling, to describe briefly the organisms involved and to present the biological basis for the use of chlorination for mussel fouling control. Finally, a general introduction to chlorination chemistry is presented along with a bibliography giving references from a biologist's viewpoint.

MARINE FOULING

The flora and fauna which compose this community include bacteria, encrusting algae and the more typical seaweeds, mussels, barnacles, hydroids, worms, sea squirts and a host of free-living scavengers such as crabs and their relatives [2], [3], [4]. The organisms colonize surfaces in a recognizable order which generally starts with physical deposition of organic molecules, followed in quick succession by bacteria, which frequently exude slime as a metabolic by-product and a substrate in which to live, and then the larger organisms. Seaweeds and attaching animals all require a hard surface on which to adhere and it is this above all that industrial and cultural developments provide. Many of the animals are net or filter feeders whose growth is enhanced by the continual renewal of the water in which they are immersed. Thus it is in the use of seawater for cooling purposes in industrial plant

† This article is reproduced by permission of the CEGB. Copyright in the article remains the property of the CEGB.

where water is continually renewed by pumping that the largest impact from marine fouling occurs.

Accepting that most surfaces, with the possible exception of cupro-nickel, are vulnerable to colonization by the fouling community the following methods are available to prevent settlement and reduce and control growth. These are: physical removal (heat or mechanical) and exclusion, surface treatment and water treatment (Table 5.1).

Fouling of larger diameter pipes (1–3 m) has been readily controlled by heat treatment. In this the water is brought to a temperature of 40°C over a period of 24–36 hours and held at the required temperature for one hour. This will effectively kill all macrofouling encountered in North European waters [5]. This type of treatment requires monitors connected to the main water system so that the rate of colonization and growth can be checked to determine the time for each successive treatment. Civil engineering, which may be expensive, must be undertaken to provide cross linkages to enable recirculation of the discharge water into the intakes (Fig. 5.1). However, where used this is a very effective means of control of fouling provided good screening is available to catch the dead shells upstream of critical plant components. Against this form of control is the fact that in power stations at least it has not proven suitable for fouling control in auxiliary circuits. The only environmental consequence is a temporary release of dead biota, which will soon be removed by the scavengers of the sea bed. With regard to other physical methods, if velocities greater than $3.0\ \mathrm{ms}^{-1}$ are maintained throughout the pipework system fouling will not develop but such conditions may only be applicable in large systems such as power stations. In ships, refineries and other chemical plant demand for water is lower and pipework too small in diameter and too distant from the intake to keep up a regime of high water velocity [6]. Additionally, the costs of installing and running pumps to maintain high water velocities are probably too high. Bacterial slimes which develop in heat exchangers of the shell and tube variety can be controlled very effectively by on-load cleaning with sponge rubber balls or brushes, as, for example, in the Taprogge system (Fig. 5.2). They require adequate physical space for the injection and ball collecting equipment. Recent developments suggest that such methods may shortly be applicable to small bore pipe systems. Exclusion suggests screening and filtration, but it should be noted that most of the macrofouling organisms start life with some free-living larval stage. Such stages form part of the marine plankton and as such are too small (1.0 mm or less in maximum dimension) to be filtered out of the water for a reasonable cost.

Surfaces can be treated to discourage larval settlement; antifouling paints are a good example of surface treatments. However, these paints have a limited life and so must be renewed. For industrial plant a reasonable life is that which does not require maintenance or replacement at more frequent intervals than the maintenance intervals of the plant which is being protected. These coatings are toxic and the life-history of the toxic release curve varies with the type of paint and toxin used. Currently periods of 3 to 5 years are the maximum guaranteed protection. Since toxic paints work by releasing the active agent into the water they will not only kill the fouling but add the toxic burden to the receiving waters. Much has now been written on the environmental impact of TBTO (tributyl tin oxide) derived from antifouling

Table 5.1 — Recommended methods for antifouling control

Method		Control			Advantages	Disadvantages
		Condenser Slime	Auxiliary Circuits	Main C–W Macrofouling		
1.	Physical	No	Limited			
1.1	Heat		Application	Yes	1. All marine fouling if T°C high enough for long enough.	1. Achieved only with specific cooling water circuit design, and consideration of turbine characteristics.
	onshore intakes — recirculation offshore intakes — reverse flow			regular treatment every 6 weeks following initial treatment in springtime	2. No continuous discharge of toxic substances.	2. Discharge T°C may require special consent regulations.
					3. 100 per cent efficiency well established.	3. Very difficult to retrofit.
						4. Auxiliary circuits and condensers require separate treatment.
1.2	Mechanical				No toxic discharge — continuous cleaning.	1. Installation costs high.
1.2.1	continuous treatment (e.g. Taprogge)	Yes	No	No	1.	2. Operational costs in ball replacement may be high.
					Maximises condenser performance.	
					2.	3. Taprogge system may become fouled and need chemical or mechanical cleaning.
1.2.2	Manual cleaning	No	No	Yes	1. Cheap.	1. Labour intensive.
					2. Station shut-down only at annual overhaul.	2. Requires careful monitoring to prevent extra outrages.
					3. Small circuits with on-shore.	3. System must be dewatered.
					4. No toxic discharge.	4. Decomposition of debris causes ventilation problems for workers.
2.	Chemical					1. Safety problems with transport and storage of liquified chlorine gas.
2.1	chlorine (including gas, hypochlorite and electrolytic sources)	Yes	Yes	Yes	1. Worldwide proven efficiency.	2. Produces organo-halogen compounds in reaction with dissolved organic matter.
					2. Works for both condensers and culverts/tunnels	3. Cost efficiency relates with water quality rather than direct demand for fouling control.
					3. Rapid loss of acute toxicity within the power station.	
					4. Easy to measure TRO.	4. Variations in chlorine demand prevents the maintaining of constant TRO limits at discharge.
					5. Acceptable costs even to retrofit.	
					6. Works in any type or age of power station.	5. Acute toxic effects on plankton within power station.
2.2	Toxic paints and sheeting	No	Not applicable	Yes, but limited application	1. Very useful on intake structures not protected by heat or chlorine	1. Frequent renewal, two year intervals after manual cleaning.
					2. Sheets may have a life of 5 or more years.	2. Not all paints cleared of environmental toxicity (e.g. TBTO — see appendix).

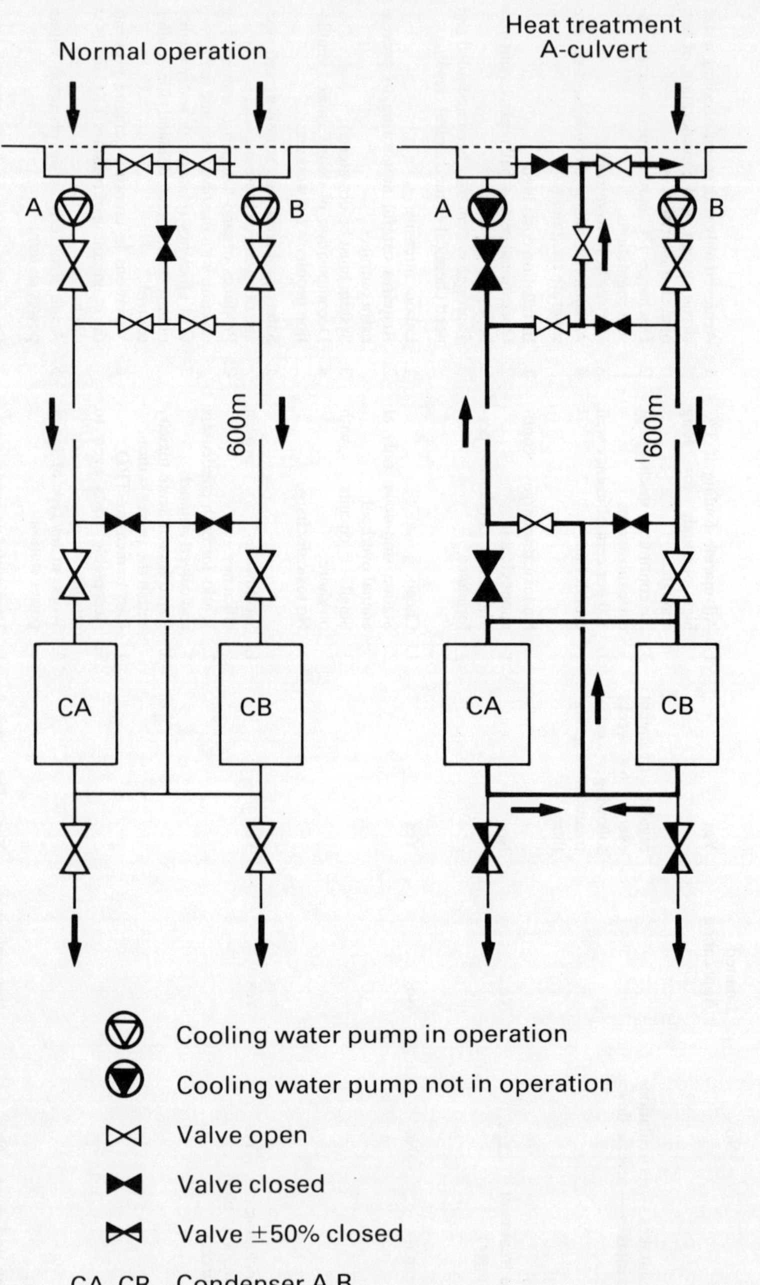

Fig. 5.1 — Lay-out of inlet cooling water system showing valves and cross-connections needed for thermal control of macrofouling.

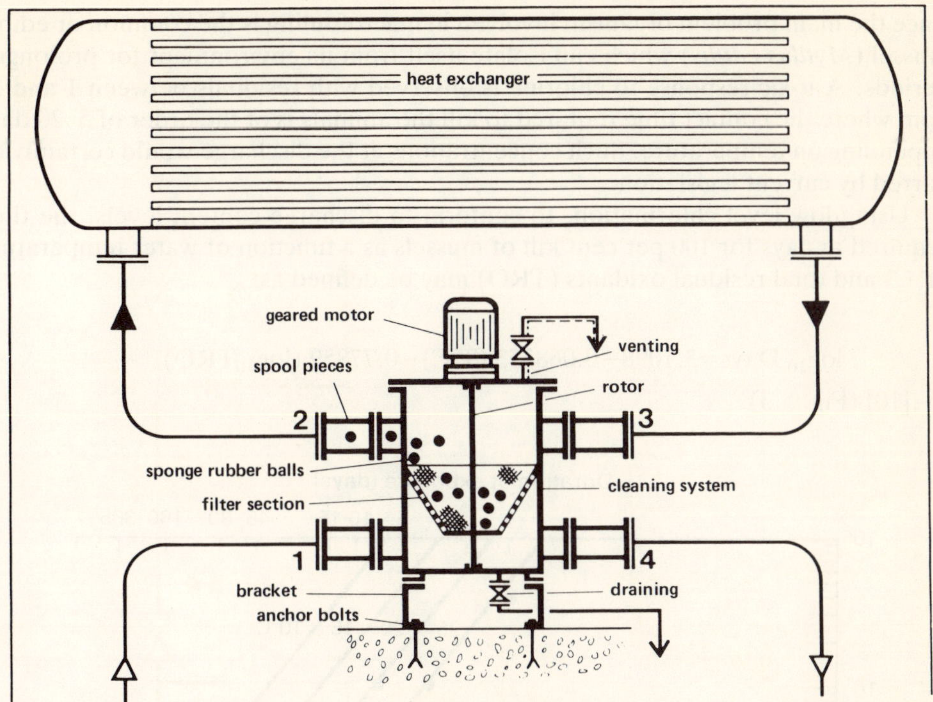

Fig. 5.2 — An on-line sponge ball cleaning system for control of slime in shell and tube condensers.

usage [7]. Pipes can also be lined with ceramics, etc., but once the integrity of the lining is breached, corrosion of the underlying pipe becomes a serious problem.

As a remedy for all forms of fouling (condenser slime, auxiliary and macrofouling) this then leaves us with the use of applied chemicals, which in liquid form are easy to dose and control to give the required toxic concentration. From a wide range of possible candidates [8] chlorine in one of its forms, bulk liquid, bulk hypochlorite or electrolytic on-site generation of hypochlorite, is the most widely used and best understood.

For slime control in marine condensers, shock doses of 2–$10\ \mathrm{mg\ l^{-1}}$ at the point of injection, which should be as close as possible to the zone to be treated, for 10–20 minutes every 4 to 8 hours are used. In effect, the hypochlorite solution denatures the slime but does not kill the bacterial cells. As it is the increase in wall thickness largely caused by the slime which reduces heat transfer, shock treatments are adequate to maintain condenser performance. Such regimes can be optimized for minimum chlorine usage and recently it has been shown that very low doses of hypochlorite ($\leqslant 0.02\ \mathrm{mg\ l^{-1}}$ TRO$\equiv$Total chlorine+any other active halogen) given as doses three times per day will control condenser slimes. To achieve this the toxicant must be targetted at the tubes [9].

Such intermittent chlorination is not appropriate for the control of macrofouling since the main problem organism involved in macrofouling is the common or edible mussel (*Mytilus edulis*) which can isolate itself from its environment for prolonged periods. A toxic response to chlorine is observed with residuals between 1 and 10 ppm where the contact time required to kill the animals is of the order of 5–20 days depending on temperature. Such concentrations at the discharge would certainly be barred by current legislation.

Using low level chlorination, to conform to discharge content levels, the time required in days for 100 per cent kill of mussels as a function of water temperature (T°C) and total residual oxidants (TRO) may be defined as:

$$\log_{10} \text{Days} = 3.1098 - 0.068353 \, (T°C) - 0.77859 \, (\log_{10}\text{TRO})$$

[10] (Fig. 5.3).

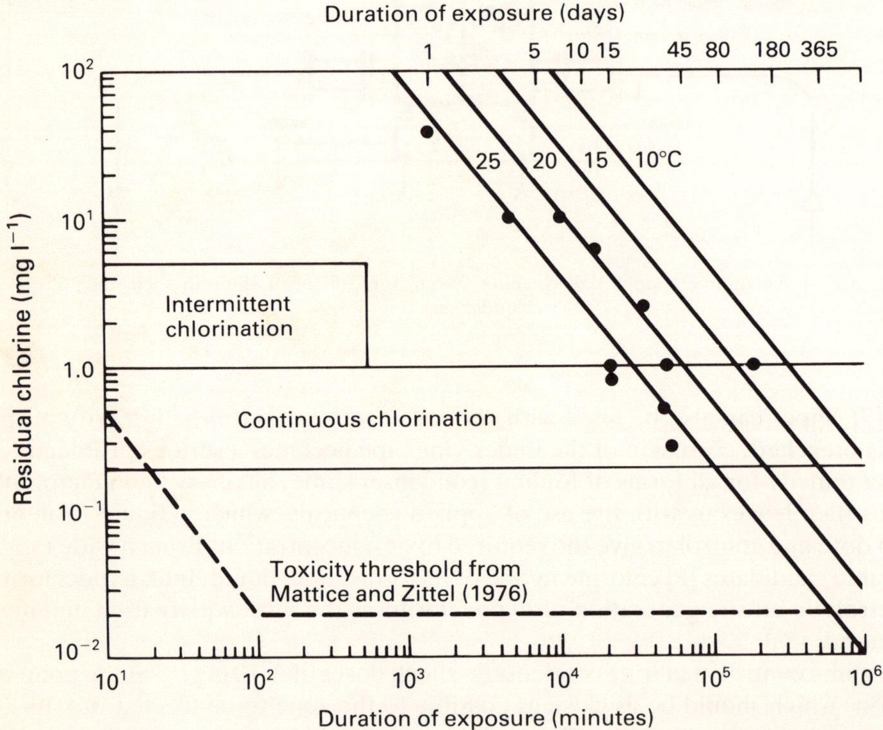

Fig. 5.3 — Time to kill off mussels using low level continuous chlorination based on European and North American data [10].

In the arithmetic form of this equation it can be seen that there is a multiplicative effect between temperature and residual. The result is when TRO is below $1.0 \, \text{mg} \, \text{l}^{-1}$ temperature becomes the driving parameter and at natural seawater temperatures found in this country days to 100 per cent kill are: $10°C=788$; $15°C=382$; $20°C=184$; $25°C=81$. So at the natural temperatures found and using low level chlorination to avoid the chance of exceeding discharge consent levels, the equipment used to dose the chlorine must be capable of continuous functioning for long periods. Any breaks in the dose of about 48 hours or more very quickly enable mussels to repair the physiological damage and treatment must be thus considered only to have started with recommencement of dosing.

Since most fouling organisms have a colonization phase in their life-cycles, which in the case of the mussel is the post-larval or plantigrade stage, it might be possible to prevent colonization by the use of applied chemicals. In the case of chlorine this has not proved possible, as settlements take place under both continuous and intermittent dosing regimes [11]. However, it is possible to detect settlements as they occur or at worst immediately after they have occurred, by using a side stream monitor. The need to chlorinate can thus be accurately determined and the efficacy of the treatment regime can be followed. This offers a route to reduce the amounts of biocide used for the control of macrofouling. However, such an approach may conflict with the need to maintain slime-free heat exchanger surfaces when they are not maintained in good order by on-line mechanical methods. It is evident from this discussion that no single form of treatment will control all forms of fouling without significantly affecting the environment as would be the case with high levels of chlorine.

CHLORINATION CHEMISTRY

It remains therefore in the rest of this chapter to outline the chemistry of chlorination of seawater. This may be regarded as a complex extension of the fresh water story [12], [13]. The process starts with the hydrolysis of chlorine:

$$Cl_2 + H_2O \rightleftharpoons HOCl + HCl$$

$$HCl + HOCl \rightleftharpoons 2H^+ + OCl^- + Cl^-$$

which is a pH dependent equilibrium. This reaction is virtually instantaneous.

Natural waters contain nitrogen compounds, such as ammonia and nitrate, which react with hypochlorite to form various chloramine compounds

$$NH_3 + HClO = NH_2Cl + H_2O$$

$$NH_2Cl + HClO = NHCl_2 + H_2O$$

where the two by-products are in equilibrium

$$2NH_2Cl + H^+ \rightleftharpoons NH_4^+ + NHCl_2$$

dependent again on pH and also the chlorine-to-ammonia ratio. In seawater these reactions may be regarded as bromine reactions since seawater contains 68 ppm bromide and the reaction:

$$Cl_2 + 2Br^- = 2Cl^- + Br_2$$

goes to completion. In practice the rapid hydrolysis of chlorine in water would exclude molecular chlorine in seawater, but the hydrolysis products HClO and ClO$^-$ will release bromine:

$$HOCl + Br^- \rightarrow HOBr + Cl^-$$
$$ClO^- + Br^- = BrO^- + Cl^-$$
$$BrO^- + H^+ = HBrO \ .$$

As in fresh water, the hypobromous acid and hypobromite ion react with ammonia to give bromammonium compounds. Again the reaction products are pH and temperature dependent. Finally, more complex organo-amine compounds may be halogenated giving rise to THMs (trihalomethanes; e.g. chloroform, bromoform), and a host of other compounds, which are but slowly degraded and may even accumulate through food-webs. The range of reactions is very wide and a simple but useful description of seawater chlorination chemistry may be found in [13].

Finally a short bibliography of work on marine fouling organisms, their control, the chemistry of chlorination and the environmental impact of chlorination and its byproducts is appended.

ACKNOWLEDGEMENT

The work was carried out in part at the Laboratories of Technology Planning and Research Division and the paper is published with permission of the Central Electricity Generating Board.

REFERENCES

[1] Ray, D. L., *Marine Boring and Fouling Organisms*, University of Washington Press, Seattle, vol. 1, p. 536, 1959.

[2] OECD, *Catalogue of Marine Fouling Organisms*, Vol. 1, Barnades OECD, Paris, 1, p. 46, 1963.

[3] OECD, *Catalogue of Marine Fouling organisms*, Vol. 2, Polyzoa OECD, Paris, 1, p. 83, 1965.

[4] OECD, *Catalogue of Marine Fouling Organisms*, Vol. 3, Serpulids, OECD, Paris, 1, p. 79, 1967.

[5] Jenner, H. A., *Trib. CEBEDEAU*, **35**, 287, 291, 1980.

[6] Trotmann, D. W., *Report No. 348*, British Ship Research Association, 1972.

[7] Whitehouse, J. W., Khalanski, M., Saroglia, M. G. and Jenner, H. A., The

control of biofouling in Marine and Estuarine Power Stations, CEGB/EDF/
ENEL/KEMA, 1, 48, 1985.

[8] Coughlan, J. and Whitehouse, J. W., *Chesapeake Science*, **18** (1), 102, 111, 1977.

[9] EPRI, Condenser-targeted chlorination design, *Rept. No. EPRI-CS*4279, Palo Alto, California, 1, 274, 1985.

[10] Lewis, B. G., CEGB Report, *TPRD/L/2810/R85*, 1, 64, 1985.

[11] Jenner, H. A., in *On Condenser Macrofouling Control Technologies*, *The State-of-the-Art*, Symposium 1–3 Hyannis, Ma., USA, EPRI, June 1983.

[12] White, G. C., *Handbook of Chlorination*, Van Nostrand Rheinhold Co., New York, Vol. 1, p. 744, 1972.

[13] Goodman, P. D., *Br. Corros. J.*, **22** (1), 56, 62, 1982.

BIBLIOGRAPHY

Fouling and Its Control

Ambrogi, R., Saroglia, M. G., and Scarano, G., 1982, Water treatment for fouling prevention. Residual toxicity of an organo-tin compound towards marine organisms. Communication in *Proceedings of the 9th Annual Aquatic Toxicity Workshop*, Edmonton, Alberta, Nov. 1982, pp. 164–170.

Blok, J. W. and de Geelan, H. J. F. M., 1958, The substratum required for the settling of mussels (*Mytilus edulis L.*), *Arch. Neerl. Zool.*, XIII, 1 suppl., 446–460.

Board, P. A. and Collins, T. M., 1965, Physical forces in marine fouling, *Discovery*, May 1965.

Board, P. A., 1983, The settlement of post larval *Mytilus edulis* (settlement of post larval mussels), *J. Moll. Stud.*, **49**, 53–60.

Boeitius, J., 1962, Temperature and growth in a population of *Mytilus edulis* from the Northern Harbor of Copenhagen (the sound). *Med. Komm. Dan., Fiskeri-og Havunders (n.s.)* **3**, 339–346.

Chadwick, W. L., Clark, F. S. and Fox, D. L., 1950, Thermal control of marine fouling at Redondo Steam of the Southern California Edison Company. *Transactions ASME,* **72**, 127–131.

Coughlan, J., 1970, Power stations and aquatic life, in *The Effects of Industry on the Environment* (ed. Cowell, E. B.), a Symposium held at Orielton Field Centre, Pembroke, South Wales.

Coughlan, J. and Davis, M. H., 1981, Effects of chlorination on entrained plankton at several United Kingdom coastal power stations, in *Water Chlorination: Environmental Impact and Health Effects*, Vol. 4, pt. 2 (eds. Jolley, R. L. *et al.*), pp. 1053–1063, Ann Arbor Science Publishing, 1983.

Coughlan, J. and Davis, M. H., 1985, Concentrations of chlorine around marine cooling water outfalls: validation of a model, in *Water Chlorination: Chemistry, Environmental Impact and Health Effects*, Vol. 5 (eds. Jolley, R. L. *et al.*), pp. 1459–1468, Lewis Publishers Inc., Chelsea, Mi, USA.

Coughlan, J. and Fleming, J. W. 1978, A versatile pump sampler for live zooplank-ton, *Estuaries*, **1** (2), 132–135.

Coughlan, J. and Whitehouse, J. W., 1977, Aspects of chlorine utilization in the United Kingdom, *Chesapeake Science,* **18** (1), 102–111.

Crisp, P. J. and Ryland, R. S., 1960, Influence of filming and surface texture on the settlement of marine organisms, *Nature* **185**, 119.

Dare, P. J., 1973, The stocks of young mussels in Morecambe Bay, Lancashire, *Shell Fish Information Leaflet*, nr. 28, 14 pp, MAFF.

Davis, D. S. and White, W. R. 1966, Molluscs from a power station culvert, *J. Conch,* **26**, 33–38.

Davis, M. H. and Coughlan, J., 1977, Response of entrained plankton to low-level chlorination at a coastal power station, in *Water Chlorination: Environmental Impact and Health Effects*, Vol. 2 (eds. Jolley, R., L. *et al.*), pp. 369–376, Ann Arbor Science Publishers, 1978.

Davis, M. H. and Coughlan, J., 1981, A model of predicting chlorine concentration within marine cooling circuits and its dissipation at outfalls, in *Water Chlorination: Environmental Impact and Health Effects*, Vol. 4, pt 1 (eds. Jolley, R. L. *et al.*), pp. 347–357, Ann Arbor Science Publishers, 1983.

Delile, G., Gasparini, R., Hawes, F. B., Koops, F. and Whalberg, B., 1982, Fouling prevention in cooling water systems of electric power stations, *Report for 19th International Unipeded Congress, 1882*.

Dobson, J. G., 1945, The control of fouling organisms in fresh- and saltwater circuits. *Trans. ASME*, pp. 247–265.

Fleming, J. M. and Coughlan, J., 1978, Preservation of vitally stained zooplankton for live/dead sorting, *Estuaries*, **1** (2) 135–137.

Field, I. A., 1922, Biology and economic value of the sea mussel *Mytilus edulis*, *Bull. US Bur, Fish.*, **38**, 127–259.

Foster, B. A., 1969, Tolerance of high temperatures by some intertidal barnacles, *Marine Biology*, **4**, 326–332.

Fox, D. L. and Concoran, E. F., 1959, Thermal and osmotic countermeasures against some typical marine fouling organisms, *Corrosion*, **14**, 31–32.

Holmes, N., 1970, Marine fouling in power stations, *Mar. Poll. Bull.*, **1** (n.s.) (7), 105–106.

Holmes, N., 1970, Marine fouling in chlorinated cooling systems, *Chem. and Ind.* **39** (9), 1244–1247.

Jenner, H. A., 1980, The biology of the mussel *Mytilus edulis* in relation to fouling problems in industrial cooling water systems, *La tribune de CEBEDUAU*, **33**, 13–19.

Jenner, H. A., 1982, Physical methods in the control of mussel fouling in seawater cooling systems, *La tribune de CEBEDEAU*, **35**, 287–291.

Jenner, H. A., 1983, Control of mussel fouling in the Netherlands: experimental and existing methods, in: *On Condenser Macrofouling Control Technologies — The State-of-the-Art*, Symposium, Hyannis, Ma., June 1983.

Koops, F. B. J., 1975, Afzetting van Organisemen in Koelwartersystemen, *Elektrotechniek*, **53**, 550–559.

Lewis, B. G., 1983, Development of preliminary model to predict time in days required for 100% kill of mussels subjected to continuous chlorination, *CEGB Report No. TPRD/L/2496/N83*.

Stock, J. N. and Strachan, A. R., 1977, Heat as a marine fouling control process at coastal electric generating stations, in *Biofouling Control Procedures*, pp. 55–62 (ed. Jensen, L. D.), Marcel Dekker, New York.

Travade, F., 1980: Premiéres observations relatives au développement des moules dans les ouvrages de prise d'eau de la centrale nucléaire de Gravelines, *Rapport EDF-DER*. HE/31–80–40.

Whitehouse, J. W., Khalanski, M. and Saroglia, M. G., 1983, Marine macrofouling control experience in the UK with an overview of European parctices, in *On Condenser Macrofouling Control Technologies — The State-of-the-Art*, Symposium, Hyannis, Ma., June 1983.

Whitehouse, J. W., Khalanski, M., Saroglia, M. G. and Jenner, H. A., 1985, The control of biofouling in marine and estuarine power stations, *Joint CEGB, EDF, ENEL, KEMA report*, 48 pp.

Chlorination Chemistry and Analysis

Bousher, A., Brimblecombe, P. and Midgley, D., 1981, Reactions of chlorine residuals in saline waters, *28th Congress of Int. Union of Pure and Appl. Chem.*, Vancouver, Canada, pp. 17–22, August 1981.

Dimmock, N. A. and Midgley, D., 1979, Modified amperometric membrane probes for determining free and residual chlorine in saline cooling waters, *Water Res.*, **13**, 1317–1327.

Dimmock, N. A. and Midgley, D., 1979, Determination of residual oxidants in chlorinated sea water with voltammetric membrane electrodes, in *Water Chlorination: Environmental Impact and Health Effects*, Vol. 3 (eds. Jolley, R. L. *et al.*), pp. 263–272, Ann Arbor Science Publishers, 1980.

Dimmock, N. A. and Midgley, D., 1982, Performance of the Orion 97–70 total residual chlorine electrode at low concentrations and its application to the analysis of cooling waters, *Talanta,* **29**, 557–567.

Midgley, D., 1980, A review of methods for determining low levels of residual chlorine in fresh water and sea water, conference on *Local Generation and Use of Chlorine and Hypochlorite*, London, 14–16 Oct. 1980, Society of Chemical Industry.

Midgley, D., 1980, Monitoring residual oxidants in cooling water, in *Electroanalytical Techniques: Industrial and Chemical Applications,* 25–27 Nov. 1980, London.

Midgley, D., 1980, Chlorine disappearance in sea water, *Wat. Res.*, **14**, 1559–1560.

Midgley, D., 1981, Review of the reactions of organic compounds in chlorinated cooling water, *Power Ind. Res.*, **1**, 3–15.

Midgley, D., 1983, A bromide-selective electrode-redox electrode cell for the potentiometric determination of bromine and free residual chlorine, *Talanta,* **30** (8), 547–554.

Scarano, G. and Saroglia, M. G., 1980, Residual chlorine analysis: a comparison between amperometric and potentiometric methods, in *Condenser Biofouling Control Symposium Proceedings* (eds. Garey, J. *et al.*), Ann Arbor Science Publishers, pp. 373–381, Ann Arbor, Michigan (1980).

Scarano, G., Saroglia, M. G. and Festa, V., 1983, Potentiometric total residual 'chlorine' analysis in a continuous flow system, in *Water Chlorination, Environmental Impact and Health Effects,* Vol. 4(1). (eds. Jolley, R. L. *et al.*), pp. 761–772, Ann Arbor Science Publishers Inc., 1983.

Environmental Impact

Coughlan, J. and Davis, M. H., 1981, Effects of chlorination on entrained plankton at several United Kingdom power stations, in *Water Chlorination: Environmental Impact and Health Effects*, Vol. 4, pt 2, (eds. Jolley, R. L. *et al.*), pp. 1053–1063, Ann Arbor Science Publishers, 1983.

Coughlan, J. and Davis, M. H., 1984, Modelling chlorine dissipation from marine cooling water outfalls, *Conference on Environmental Contamination*, London, July 1984.

Dalesmont, R. and Delattre, J. M., 1982, Etude microbiologique des effets thermiques et de la chloration en Bord de Mer (Gravelines), *EdF Report No. HE/31 — 82–36*.

Davis, M. H. and Coughlan, J., 1977, Response of entrained plankton to low-level chlorination at a coastal power station, in *Water Chlorination: Environmental Impact and Health Effects*, Vol. 2 (eds. Jolley, R. L. *et al.*), pp. 369–376, Ann Arbor Science Publishers, 1978.

Davis, M. H. and Coughlan, J., 1981, A model for predicting chlorine concentration within marine cooling circuits and its dissipation at outfalls, in *Water Chlorination: Environmental Impact and Health Effects*, Vol. 4, pt 1 (eds. Jolley, R. L. *et al.*), pp. 347–357, Ann Arbor Science Publishers, 1983.

Fiquet, J. M., 1978, Contribution à l'étude du dosage du chlore dans l'eau de mer, *T.S.M. L'eau,* April, 1978, pp. 239–245.

Khalanski, M., 1981, Influence du functionnement d'une centrale thermique sur la production primaire planctonique due port de Dunkerque, *Journées de la Thermo-écologie, EDF Direction de l'Equipment*, Conference 14–15 Nov. 1979, Nantes.

Langford, T. E., 1983, *Electricity Generation and the Ecology of Natural Waters*, pp. 342, Liverpool University Press.

Mattice, J. S. and Zittel, H. E., 1976, Site-specific evaluation of power plant chlorination, *JWPCF,* **48** (10): 2284–2308.

Mix, M. C. and Schaffer, R. L., 1983, Concentrations of unsubstituted polynuclear aromatic hydrocarbons in Bay Mussels (*Mytilus edulis*) from Oregon, USA, *Mar. Envir. Res.,* **9**, 193–209.

Opresko, D. M., 1980, Review of open literature on effects of chlorine on aquatic organisms, *EPRI-EA-1491*.

Saroglia, M. G., Queirazza, G. and Scarona, G., 1980, Water quality criteria for aquaculture in thermal effluents, heavy metals and residual antifouling products, in *Proceedings World Symposium on Aquaculture in Heated Effluents and Recirculation Systems*, Stavanger 28–30 May 1980, Vol. 1, Berlin 1981.

Sugam, R. and Helz, G. R., 1977, Specification of chlorine produced oxidants in marine waters: theoretical aspects, *Chesapeake Sci.,* **18** (1): 113–118.

Vanderhost, J. R., 1982, Effects of chlorine on marine benthos, *EPRI Report No. EA-2696*, Palo Alto, Ca.

6

Copper base alloys in chlorinated waters

R. Francis
BNF Metals Technology Centre, Wantage, Oxon.

INTRODUCTION

BNF Metals Technology Centre has been active in the study of marine corrosion for over 50 years, particularly in the development and testing of copper alloys for piping and heat exchangers. During the 1970s five out of a fleet of seven super tankers began to experience a series of rapid condenser tube failures which took the form of impingement attack, especially at scratches and other points of local damage. All seven tankers were tubed in aluminium brass and ferrous sulphate dosing was used on each vessel. The five tankers experiencing corrosion problems had continuous chlorine dosing equipment while the other two vessels had none. When other companies reported similar failures, BNF undertook a research programme to investigate the possible interaction of ferrous sulphate dosing and continuous chlorination. The results showed that continuous chlorination could be extremely detrimental under some circumstances. Following this, BNF carried out further work, largely funded by the International Copper Research Association (INCRA), which determined safe operating conditions, examined the changes chlorine caused in the protective films, and demonstrated the effects of chlorine when ammonia or sulphides are also present in seawater. This chapter presents a summary of the results.

THE EFFECT OF CHLORINE ON CORROSION

The corrosion tests were carried out in once-through seawater using the Campbell condenser tube test rig [1]. The different corrosion zones in this apparatus are shown in Fig. 6.1. The four copper alloys most commonly used in heat exchangers were tested as shown in Table 6.1. In the initial work continuous chlorine levels from 0.3 ppm to 4 ppm were investigated, both with and without ferrous sulphate additions [2]. The major effect of chlorine occurred under impingement conditions,

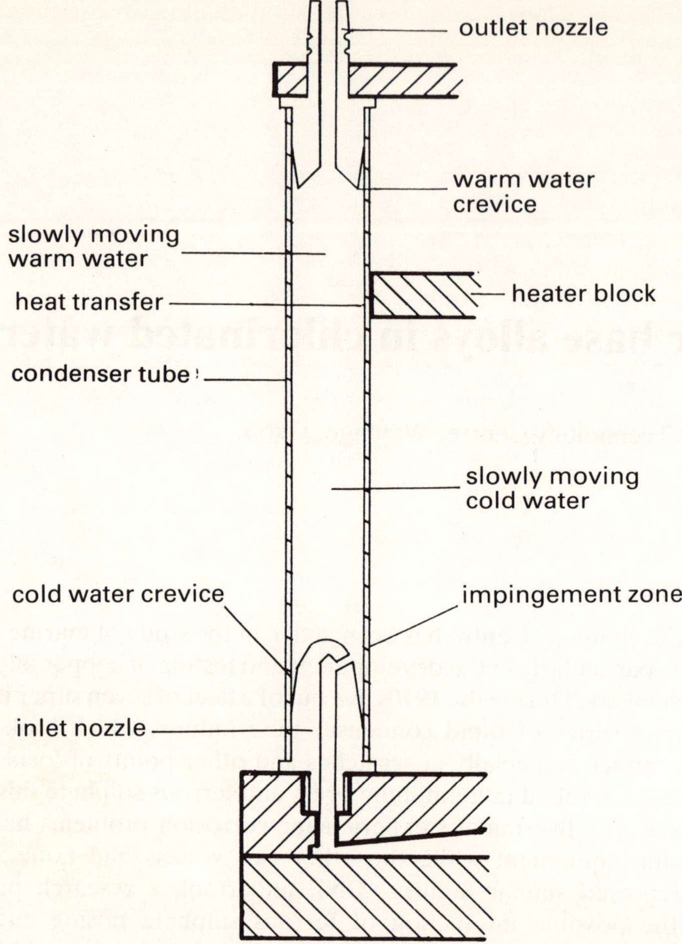

Fig. 6.1 — The different corrosion zones on the Campbell condenser tube test rig specimens.

and the results for aluminium brass showed that when the protective film broke down deep localized attack could occur. The effect of chlorine appeared to be to reduce the velocity at which impingement attack occurred, as impingement attack became more and more severe as the chlorine level was increased, as shown in Fig. 6.2. At 4 ppm chlorine some of the tubes failed in less than the 60-day period. Not only the depth of attack but also the diameter of the region suffering attack varied with chlorine concentration, as shown in Fig. 6.3. The impingement zone consists of a high velocity central zone surrounded by an area in which the turbulence decreases radially until the slow flow region higher up the tube under test is reached ($\simeq$ jet velocity $\times 10^{-2}$). Thus the increasing diameter of the attacked area with chlorine concentration shows that the corrosion film was becoming less protective, as the chlorine concentration increased. However, at 4 ppm chlorine a more protective film formed which broke

Table 6.1 — Nominal composition of alloys investigated

Alloy	Standard		Composition (Wt %)						
	British	USA	Cu	Zn	Ni	Fe	Mn	Al	As
Aluminium brass	CZ110	68700	76	22	—	—	—	2	0.04
90/10 Copper–nickel	CN102	70600	87.5	—	10	1.5	1	—	—
70/30 Copper–nickel	CN107	71500	68.5	—	30	0.5	1	—	—
Copper–nickel–iron– manganese	CN108	71640	66	—	30	2	2	—	—

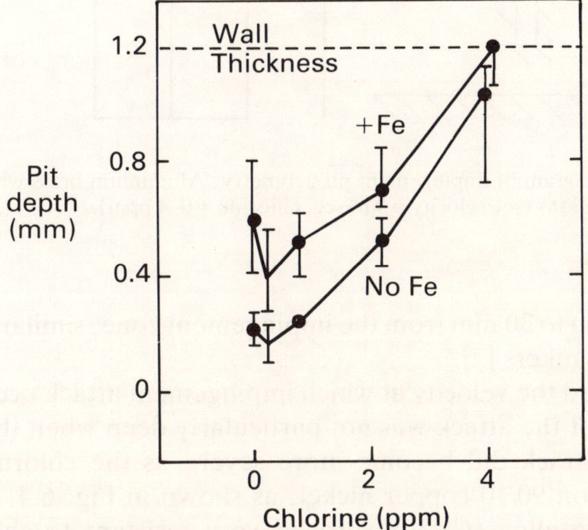

Fig. 6.2 — Depth of impingement attack against chlorine level for aluminium brass (jet velocity = 9 m/sec).

down only under the most turbulent conditions. The effect of chlorine on the nature of the protective film is discussed in more detail in the next section.

The film on all the tubes was scratched half-way through the test, and on tubes which were not receiving chlorine this healed over very rapidly. However, on tubes receiving water dosed with chlorine there were definite signs of impingement attack, and on tubes which were also dosed with ferrous sulphate there was deep attack

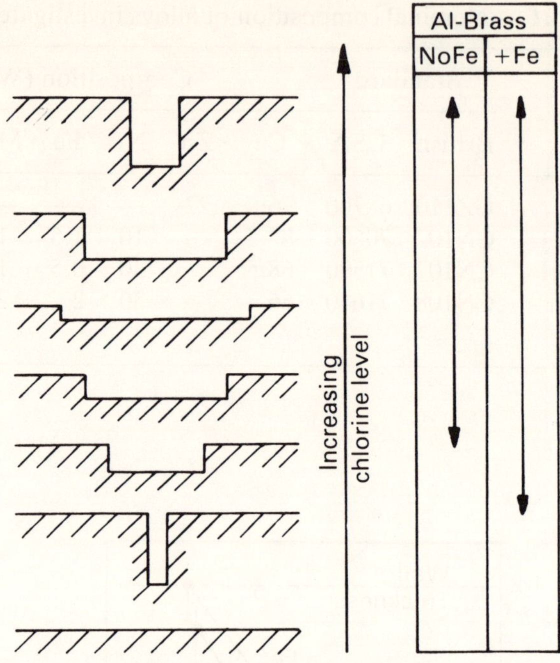

Fig. 6.3 — Schematic diagram of impingement pit geometry. Aluminium brass with chlorine additions (jet velocity = 9 m/sec; chlorine = 0–4 ppm).

forming a long groove up to 30 mm from the impingement zone, similar to the attack observed on the super tankers [2].

Chlorine also reduced the velocity at which impingement attack occurred on the copper nickel alloys, but the attack was not particularly deep when the film broke down. However, the attack did become more severe as the chlorine level was increased, particularly on 90/10 copper nickel, as shown in Fig. 6.4. The copper- –nickel–iron–manganese alloy (CN108) proved very resistant to chlorine and a significant increase in impingement corrosion was only observed at 4 ppm chlorine [2], as shown in Fig. 6.5.

One striking observation on all the alloys was the effect of chlorine when ferrous sulphate additions were also made. Ferrous sulphate ia added to seawater to improve the resistance of copper alloys, particularly aluminium brass, to impingement attack. However, it was observed that less and less hydrated iron oxide was deposited on the tubes as the chlorine level increased. Subsequent work by Effertz and Fichte [3] has shown that chlorine prevents the essential colloidal stage of FeOOH being formed when ferrous sulphate is added to seawater. Instead a floc is formed which is carried through the system, Without this iron-rich deposit aluminium brass in particular becomes more susceptible to impingement attack. For this reason it is recommended that chlorination is switched off when ferrous sulphate dosing is being carried out.

Ferrous sulphate additions are not normally used with copper nickel alloys, which

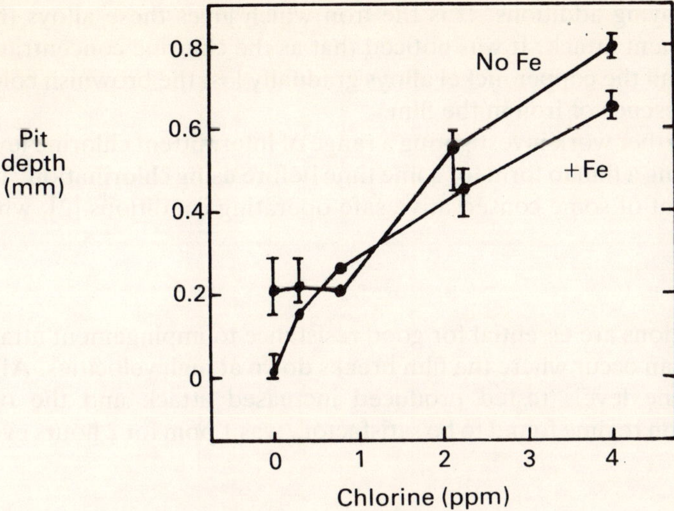

Fig. 6.4 — Depth of impingement attack against chlorine level. 90/10 copper nickel (jet velocity = 9 m/sec).

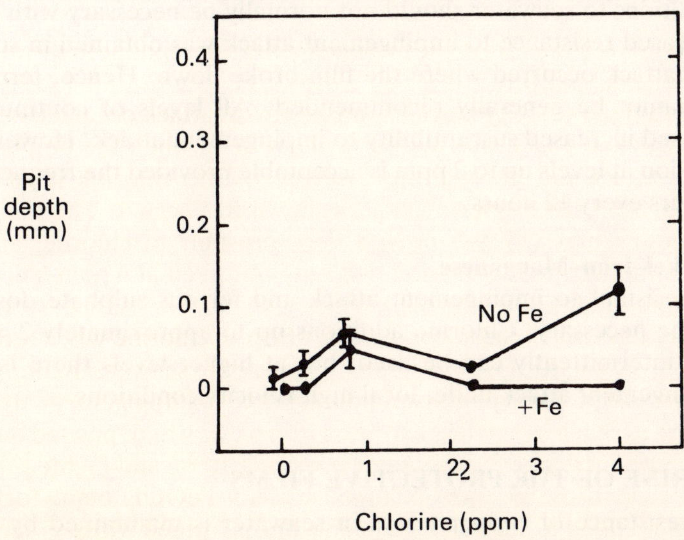

Fig. 6.5 — Depth of impingement attack against chlorine level. 66/30/2/2 Cu-Ni-Fe-Mn (jet velocity = 9 m/sec).

all contain iron as alloying additions. It is the iron which gives these alloys their resistance to impingement attack. It was noticed that as the chlorine concentration increased the films on all the copper nickel alloys gradually lost the brownish colour associated with the presence of iron in the film.

BNF carried out further work investigating a range of intermittent chlorine levels and the effect of allowing a film to form for some time before using chlorination. This led to the establishment of some conservative safe operating conditions [2], which are summarized below.

Aluminium brass

Ferrous sulphate additions are essential for good resistance to impingement attack, although deep attack can occur where the film breaks down at high velocities. All of the continuous chlorine levels tested produced increased attack and the only intermittent chlorination regime found to be satisfactory was 1 ppm for 2 hours every 12 hours.

90/10 Copper–Nickel

Ferrous sulphate additions to the seawater can give increased resistance to impingement attack with this alloy, although they should not normally be necessary. Continuous chlorine dosing up to 0.3 ppm chlorine may be used but there is an increasing risk of impingement attack at higher levels. Intermittent chlorination is also acceptable, but it is more important to keep the level of chlorine dosing low (<0.5 ppm) than to control the frequency of dosing.

70/30 Copper–Nickel

Ferrous sulphate additions to seawater should not normally be necessary with this alloy. Although increased resistance to impingement attack was obtained in some tests, in others deep attack occurred where the film broke down. Hence, ferrous sulphate additions cannot be generally recommended. All levels of continuous chlorine addition caused increased susceptibility to impingement attack. However, intermittent chlorination at levels up to 2 ppm is acceptable provided the frequency does not exceed 2 hours every 12 hours.

60/30/2/2 Copper–Nickel–Iron–Manganese

This alloy is highly resistant to impingement attack and ferrous sulphate dosing would not normally be necessary, Chlorine additions up to approximately 2 ppm either continuous or intermittently can be used, but at higher levels there is an increased risk of impingement attack under local high velocity conditions.

EFFECT OF CHLORINE OF THE PROTECTIVE FILMS

The high corrosion resistance of copper alloys in seawater is maintained by the formation of a thin protective film during the first few months of service. The results obtained from the tests described above indicated that chlorine was not only preventing the formation of iron-rich deposits, but it was also altering the nature of the protective film. Tests were carried out using a recirculating seawater loop to determine the electrochemical properties of the films that form on both aluminium brass and 90/10 copper nickel in both clean seawater and seawater dosed with chlorine [2,4]. In addition, some of the tubes from the once-through seawater tests were brought back to the laboratory for testing.

The results showed that in seawater containing chlorine the potentials of both alloys moved some 100 mV electropositively, compared to those of tubes in clean seawater, within a few days of exposure. Some typical potential time curves for aluminium brass are shown in Fig. 6.6. Polarization curves showed that in the

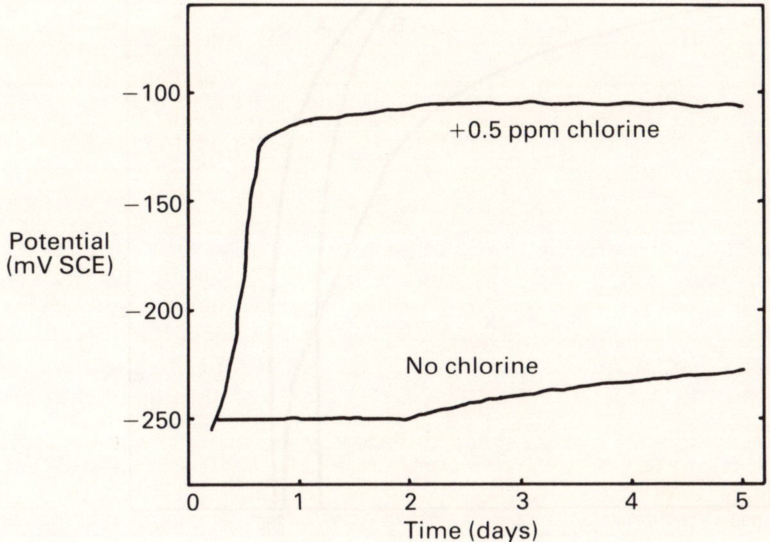

Fig. 6.6 — Potential time curves for aluminium brass.

presence of chlorine the films polarized less readily in the anodic direction. However, in the presence of chlorine increased cathodic polarization was observed compared to tubes exposed in clean seawater, which became more marked as the chlorine level increased. Some typical cathodic polarization curves for 90/10 copper nickel are shown in Fig. 6.7. At 4 ppm chlorine the films appeared totally different to those formed in clean seawater, and showed a very marked degree of cathodic polarization. The films formed at this chlorine level were essentially composed of copper and chloride, in marked contrast to the films which form in clean seawater. It was concluded [4] that in the presence of chlorine the film is changed from the normal composition to one rich in copper and chloride ions which is capable of inhibiting the cathodic reaction. Although this film produces low general corrosion rates it has a low resistance to mechanical damage, and because iron containing films do not form at high chlorine levels it would be more susceptible to impingement attack than the film which forms in the absence of chlorine. This mechanism resolves the conflicting results published on the effects of chlorine on the corrosion of copper alloys. Some authors testing under slow flow conditions [5,6] have reported lower corrosion rates

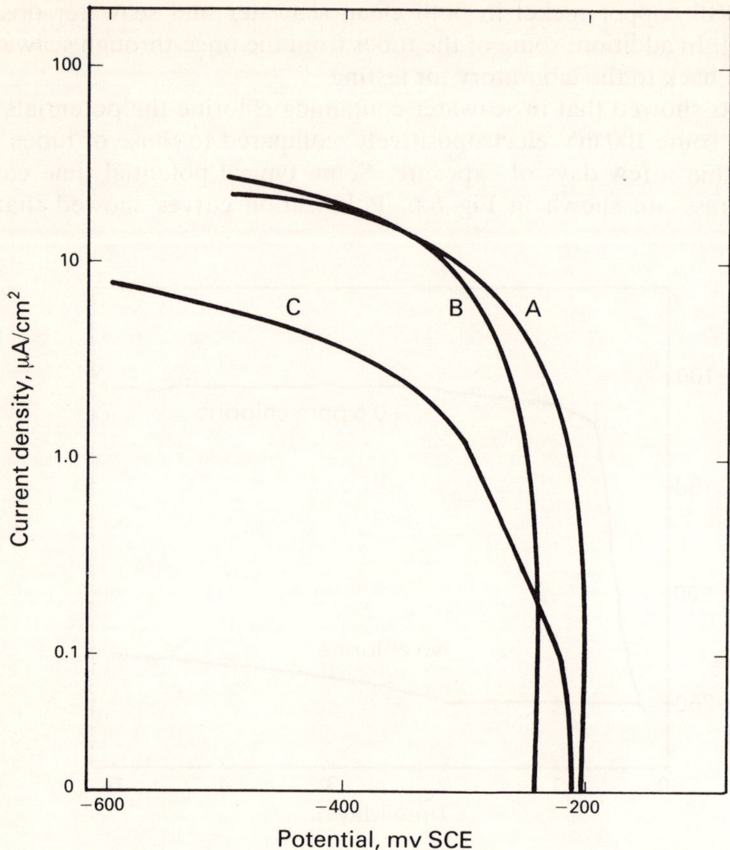

Fig. 6.7 — Cathodic polarization curves for 90/10 copper nickel after 2 months' test: A, no
additions; B, 0.5 ppm chlorine; C, 4 ppm chlorine.

in the presence of chlorine, while other work at higher velocities [2,7] has shown
increased corrosion.

EFFECT OF CHLORINE AND POLLUTION

There are many places throughout the world where seawater contains pollutants.
The most common are ammonia and sulphide (usually present as hydrogen sulphide)
and BNF has examined the possible interactions between chlorine and both ammonia and sulphide. In addition BNF has carried out corrosion tests over a limited range
of concentrations with the Campbell Condenser tube test rig. The results [8,9]
showed that in some circumstances the presence of chlorine can reduce the deleterious effects of ammonia [8] and sulphide [9] but the effects varied from alloy to alloy,
and in some circumstances severe acceleration of corrosion occurred.

For example, Figs 6.8 and 6.9 show the effect of chlorine and sulphide on

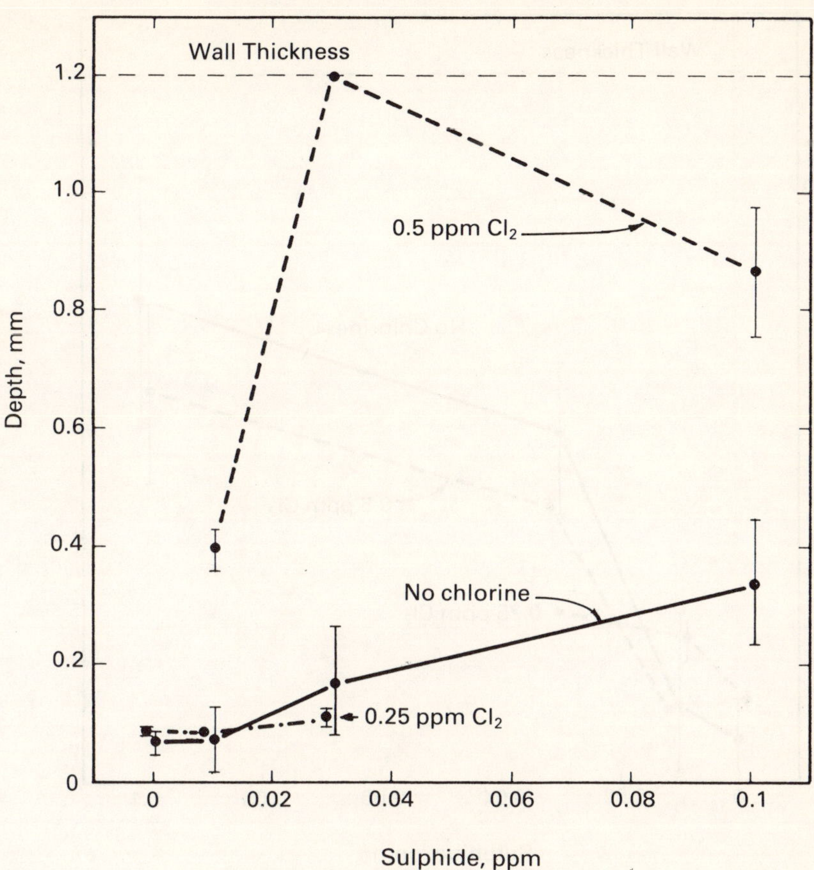

Fig. 6.8 — Depth of impingement zone attack on 90/10 copper–nickel as a function of chlorine and sulphide levels (jet velocity = 7 m/sec).

impingement attack on 90/10 and 70/30 copper nickel respectively. It can clearly be seen that 0.25 ppm chlorine had little effect on the depth of attack on 90/10 copper nickel, compared to that on tubes receiving seawater with sulphide only. At 0.5 ppm chlorine very severe attack occurred, leading to perforation of some tubes. With 70/30 copper nickel there was a substantial reduction in the depth of attack at 0.25 ppm chlorine compared to tubes receiving water containing only sulphide. With 0.5 ppm chlorine the depth of attack increased compared to that at 0.25 ppm chlorine, but it was still somewhat lower than that in tubes receiving no chlorine. The reason for this is thought to be connected with the very good resistance of 70/30 copper nickel to sulphide oxidation products. However, as the chlorine level increased the excess chlorine over that required to react with the sulphide increased, and this led to the increased attack with 0.5 ppm chlorine. The increase in attack with 0.5 ppm chlorine was much more marked with 90/10 copper nickel because this alloy does not have such good resistance to sulphide oxidation products as 70/30 copper nickel [9].

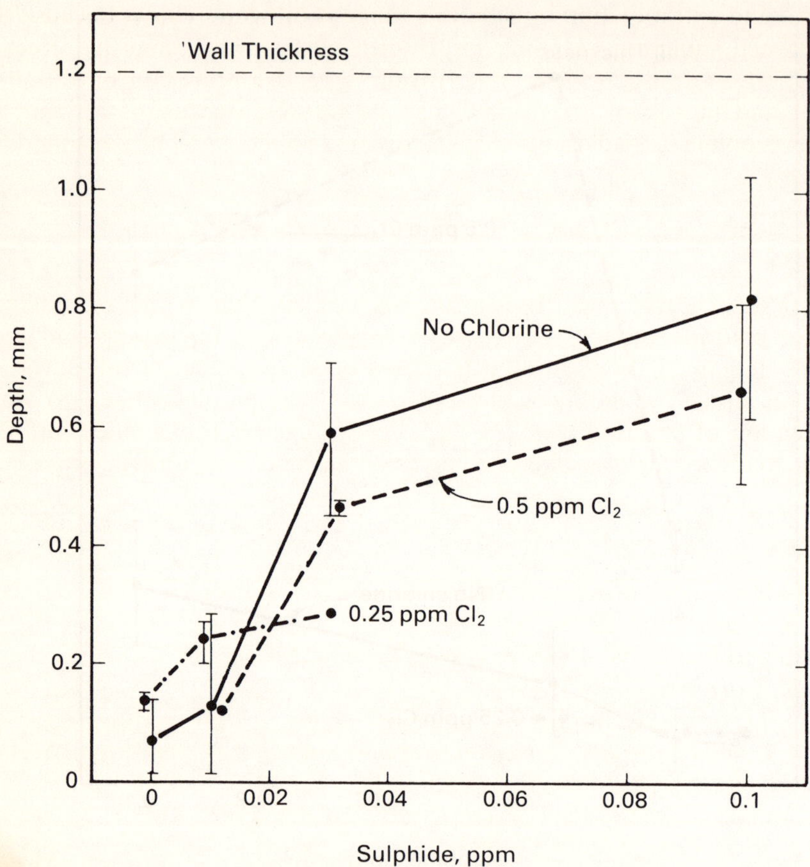

Fig. 6.9 — Depth of impingement zone attack on 70/30 copper–nickel as a function of chlorine and sulphide levels (jet velocity = 7 m/sec).

It was concluded that small chlorine additions could reduce the acceleration of corrosion of copper alloys caused by the presence of sulphides. However, chlorine concentrations greatly in excess of that required to oxidize the sulphide were to be avoided. The exception to this was copper–nickel–iron–manganese (CN108) which suffered greatly increased attack at all chlorine levels when sulphide was also present, compared to tubes exposed to water containing sulphide only.

Pollutants were also observed to affect forms of corrosion other than impingement attack. For example, 2 ppm ammonia was observed to cause a type of pitting under conditions of very slow flow and heat transfer. The presence of redeposited copper in the corrosion products made the attack appear more akin to hot spot attack than crevice corrosion. The addition of 0.5 ppm chlorine completely suppressed this type of attack on all the copper alloys tested [8].

Ammonia, sulphide and chlorine were observed to have other effects on corrosion. However, tests were only conducted over a limited concentration range of each chemical and extrapolation outside this range is difficult. This is because the reactions between ammonia, sulphide and chlorine are complex, and the effect of any combination on different types of corrosion depends on the concentration of each reactant, and the mixing time prior to entering the tube. Because of the complex nature of this problem it is advisable that the original papers [8,9] be consulted for further details.

CONCLUSION

The tests carried out at BNF over the last 10 years have enabled an understanding of the effects of chlorine on the corrosion of copper alloys for piping and heat exchangers to be gained. In addition it has been possible to specify safe levels of chlorination for all the commonly used copper alloys. The problems likely to occur when chlorination of polluted water is carried out have been examined and the combinations likely to cause accelerated corrosion have been identified.

REFERENCES

[1] Campbell, H. S. *BNF Miscellaneous Publication 577*, February 1973.
[2] Francis, R. *Mater, Perf.* **21** (8), 44, 1982.
[3] Effertz, P. and Fichte, W. *VGB Kraftwerkstechnik* **57**, 166, 1977.
[4] Francis, R. *Corr. Sci.* **26**, 205, 1986.
[5] Anderson, D. B. and Richards, B. R. *J. Engineering for Power*, July 1966.
[6] Campbell, I. and Searle, N. K. Conference on Fouling and Corrosion of Metals in Seawater, SMBA Dunstaffnage, 1982.
[7] Sato, S. *Sumitomo Light Metals Tech. Rep.*, **3** (3) 106, 1962.
[8] Francis, R. *Brit. Corr. J.* **20**, 157 1985.
[9] Francis, R. *Brit. Corr. J.* **20**, 175 1985.

7

Stainless steels in chlorinated waters

P. Hagenfeldt
AB Sandvik Steel, 45-STRP, 811 81 Sandviken, Sweden

INTRODUCTION

In recent years the use of highly alloyed stainless steels for seawater applications, e.g. heat exchangers, condensers, desalination plants, fire fighting systems, etc., has increased rapidly.

Under normal conditions biofouling will occur on stainless steel when exposed to seawater. In most applications this biofouling is undesirable and has to be removed. The most common methods for removing biofouling are chlorination or chlorination combined with mechanical cleaning. Unfortunately, the corrosivity of seawater, which is itself already a very corrosive medium, increases when chlorinated.

A literature survey by Dahl [1] in 1983, updated in 1985 [2], indicates that very little research has been conducted into the behaviour of highly alloyed stainless steels in chlorinated sea-water. This is perhaps not surprising bearing in mind that many of the alloys have only recently been developed.

AB Sandvik Steel has been involved in two independent research programmes concerned with stainless steels in chlorinated seawater. These programmes are as follows.

1. Basic investigations of highly alloyed stainless steels in seawater, with the objective of studying the effect of biofouling and chlorination on the electrochemical properties of stainless steels in natural seawater. The project has been financed by NTNF, Avesta AB and AB Sandvik Steel with the experimental work conducted by SINTEF at Trondheim, Norway.
2. Corrosion in chlorinated North Sea Water. Highly alloyed stainless steels have been exposed to continuously chlorinated ($2\,\mathrm{mg\,l^{-1}}\,\mathrm{Cl_2}$) North Sea water at 35°C using test samples that simulate practical conditions. This project has been financed by the Thermal Engineering Research Association (SVF) and the Nordic Liaison Committee for Atomic Energy (NKA). The experimental work

was conducted at Det Norske Veritas (Bergen, Norway), the Swedish Corrosion Institute and Avesta AB.

This chapter will present selected results from these research programmes.

THE EFFECT OF BIOFOULING AND CHLORINATION ON THE ELECTROCHEMICAL PROPERTIES OF STAINLESS STEELS

The performance of stainless steels in chloride solutions is highly dependent on the corrosion potential of the steel or, in other words, the redox potential of the solution. Of course, temperature, pH and chloride concentration are also important factors and will, in fact, affect the corrosion potential.

For example, Type 316 stainless steel may perform well in deaerated seawater where the combination of pH, temperature and chloride content together with a low corrosion potential is not sufficiently aggressive to cause local breakdown of the protective oxide layer. If oxygen is added to the seawater then the corrosion potential of the steel will probably rise to a value above that of the critical potential for pitting of Type 316 and localized corrosion will occur.

The susceptibility of stainless steel to pitting and crevice corrosion at different potentials in chloride solutions may be determined by measuring the Critical Pitting Temperature (CPT) and/or Critical Crevice Temperature (CCT) with a method developed at Sandvik [3].

In this test the samples, with and without applied crevice, are polarized to a predetermined level for 15 minutes at a start temperature of, usually, 20°C. If no localized attack develops in this period the polarizing current is switched off and the temperature is raised by 5°C. The samples are then polarized again for another 15 minutes to the previous value. This procedure is continued until the current density, which is continuously recorded, exceeds a critical level, indicating that the sample is corroding actively. The temperature at which the current increases with specimens without crevices is referred to as the CPT and with crevices as the CCT. In both cases the values are referred to the actual environment in which the sample was polarized.

Fig. 7.1 demonstrates the typical effect of applied potential on the critical pitting and crevice temperatures for a stainless steel tested in a chloride solution. If the corrosion potential of the steel, as represented by the applied potential in this case, increases then the CPT and CCT decrease.

These results emphasize the importance of investigating the corrosion potential of freely exposed stainless steels in seawater since both biofouling and, especially, chlorination strongly influence this parameter.

EXPOSURE TRIALS IN SEAWATER

The experiments were conducted in a circulation loop and in a PVC box. The circulation loop consisted of three parallel PVC pipes with inner diameter 110 mm connected to a common reservoir. In each of the pipes the velocity could be varied in the range 0–5 ms^{-1}. In the PVC box with a total volume of 60 litres the seawater was nearly stagnant and the flow rate at the inlet was adjusted so that the contents were

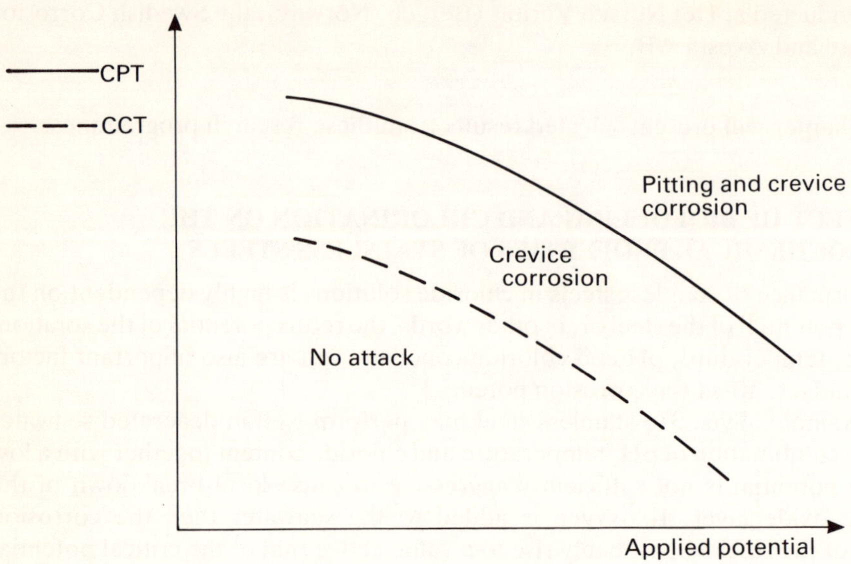

Fig. 7.1 — The effect of applied potential on critical pitting and crevice temperatures for a stainless steel in chloride solutions.

replaced every half hour. Fig. 7.2 is a schematic view of the box with some specimens in place.

Three different high-alloy stainless steels, 254SMO, Sanicro 28 and SAF 2205 were exposed to natural seawater in the circulation loop. The chemical compositions of these steels is given in Table 7.1. During the test the potentials of freely exposed specimens were measured under the following conditions [4,5,6]:

(a) seawater from <5 m depth;
(b) seawater from 60 m depth;
(c) seawater from 60 m depth without access to light;
(d) seawater from 60 m depth at a flow rate of 0.5, 1.2 and 4.5 ms^{-1};
(e) seawater from 60 m depth with intermittent chlorination.

The seawater was taken from the Trondheimsfjord or Byjford (Bergen). The temperature of the seawater was 8–10°C for all tests. Table 7.2 shows the chemical composition of the seawater from <5 m and 60 m taken from the Trondheimsfjord.

For conditions (a), (b) and (c) all three alloys were included; in conditions (d) and (e) only 254SMO and SAF 2205 were included.

The chlorination was carried out by adding sodium hypochlorite, NaClO, (10 per cent) at an amount of 10 ppm (by weight) after 31 days and 51 days of the test. Before the addition the inlet was closed and then chlorination continued for 30 minutes. After chlorination the seawater in the PVC box was exchanged 60 times during a

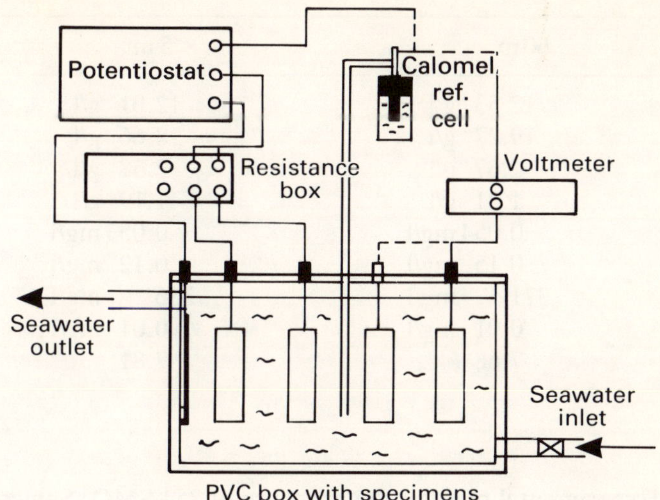

Fig. 7.2 — Schematic view of a PVC box with specimens mounted.

Table 7.1 — Nominal composition of the stainless steels tested

Grade	Weight%						Micro-structure
	C	Cr	Ni	Mo	Cu	N	
Type 316	≤0.05	17.0	11.5	2.7	—	—	Austenitic
Type 904L	≤0.02	20.0	25.0	4.5	1.5	—	Austenitic
254SMO	≤0.02	20.0	18.0	6.1	0.7	0.2	Austenitic
Sanicro 28	≤0.02	27.0	31.0	3.5	1.0	—	Austenitic
Monit	≤0.025	25.0	4.0	4.0	—	—	Ferritic
SAF 2205	≤0.03	22.0	5.5	3.0	—	0.15	Duplex

period of 90 minutes by pumping fresh water in through the inlet at a high flow rate.

For comparison, SAF 2205 was exposed to artificial seawater, prepared according to ASTM D1141-52 for 55 days at 10°C. Table 7.3 summarizes the experimental plan.

The specimens (100×50 mm) were cut from mill-annealed plate of thickness 3–5 mm. Before exposure the specimens were cleaned with soap and water, pickled in 15–20 per cent $HNO_3 + 4 - 5$ per cent HF for 20–30 minutes at room temperature,

Table 7.2 — Chemical composition of seawater from the Trondheimfjord

ion	60 m	<5 m
Na	12.53 g/l	12.01 g/l
Cl	19.27 g/l	18.66 g/l
SO$_4$	2.67 g/l	2.58 g/l
Mg	1.21 g/l	1.19 g/l
Ca	0.054 mg/l	0.053 mg/l
Fe	0.15 mg/l	0.12 mg/l
K	371 mg/l	356 mg/l
Zn	0.01 mg/l	0.01 mg/l
pH	7.68	7.81

Table 7.3 — Experimental plan for the exposure of 254 SMO, Sanicro 28 and SAF 2205

	254 SMO	SAF 2205	Sanicro 28
Seawater from <5 m	×	×	×
Seawater from 60 m	×	×	×
Seawater from 60 m without access to light	×	×	×
Seawater from 60 m. Flow rates 0.5, 1.2 and 4.5 m/s	×	×	
Seawater from 60 m with intermittent chlorination	×	×	
Artificial seawater (ASTM D 1141-52)		×	

washed with tapwater and finally exposed to air for one week. The roughness of the surface was kept in the as-received condition.

Extensive measurements were also made on cathodically polarized specimens and the results have been mentioned elsewhere [4,5,6].

The specimens were mounted in special holders in the circulation loop and with thin platinum wires in the PVC box.

RESULTS AND DISCUSSION

The corrosion potentials* of freshly exposed 254SMO, Sanicro 28 and SAF 2205 in seawater from <5 m depth with and without access to light, developed virtually identically in the respective environments.

*All potential values are vs. SCE.

Fig. 7.3 shows how the corrosion potential for Sanicro 28 developed during the 55 days of exposure. The sudden drop of the potential after 34 days of exposure in the seawater from 60 m depth was caused by a blockage in the outlet which caused the water level to rise and disturb the potential measurements. As mentioned above, the corrosion potential was not affected by alloy composition and so the curves for the other two alloys are not shown.

However, there were small differences between the different environments. The following points were noted.

1. After two days a visible slime layer built up on the surface of stainless steel in all three environments
2. The properties of this slime layer caused the corrosion potentials to increase from about -100 to $0\,mV$ to a stable value of about 300–$330\,mV$ after 20–25 days in seawater from 60 m depth with or without access to light
3. The potentials of specimens in seawater from $<5\,m$ rose more slowly and reached a value of 250–$280\,mV$ after 55 days.

The effect of biological activity (biofouling) is clearly illustrated in Fig. 7.4. The potential of SAF2205 in artificial seawater reaches a stable value of about $130\,mV$ after 10 days, which is approximately $200\,mV$ lower than in natural seawater.

Fig. 7.5 shows how the corrosion potential developed for 254SMO and SAF 2205 specimens in seawater at different flow rates during an exposure of 280 days. Again there is no difference in potential among the alloys. The curves obtained at different flow rates are similar to those obtained in stagnant seawater. The potentials of specimens exposed to flow rates of $4.5\,ms^{-1}$ appeared to be slightly higher (by 30–$50\,mV$) than those for $0.5\,ms^{-1}$. The potentials after 280 days were between 370 and $400\,mV$.

Before chlorination biofouling was allowed to develop for 31 days during which time the potentials reached stable values of about $300\,mV$. Fig. 7.6 shows the effect of 30 minutes chlorination on the potential. (The sudden drop in potential after 11 days was due to a power failure which caused the specimens to be exposed to air for approximately 12 hours.) The potentials on chlorination increased almost instantaneously to about 450–$500\,mV$ but then fell as the chlorine was washed away to about $150\,mV$ within 10 hours. This indicated that the biological activity had been 'killed'. The potentials then dropped to values corresponding to those in artificial seawater. Figs 7.7 and 7.8 show the potential behaviour during chlorination in more detail. After the potential fell to $150\,mV$ it started to increase again and reached $300\,mV$ within 20 days.

The effect of chlorination after 51 days' exposure was similar to that after 31 days.

No biofouling could be identified on the specimens that has been intermittently chlorinated after 31 and 51 days whereas biological growths could be identified on all other specimens.

The results clearly show that for stainless steel freely exposed in seawater there are three corrosion potential levels to consider:

1. $150\,mV$ (no biofouling);

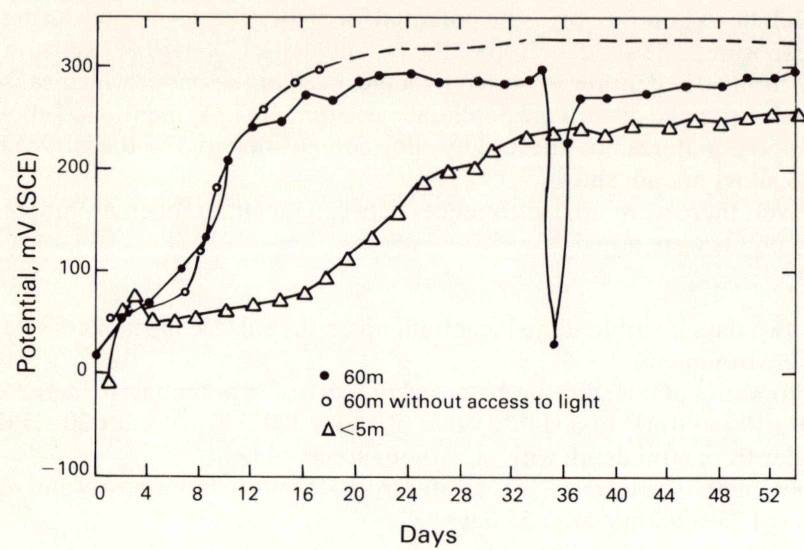

Fig. 7.3 — Corrosion potential for Sanicro 28 in seawater under three different conditions.

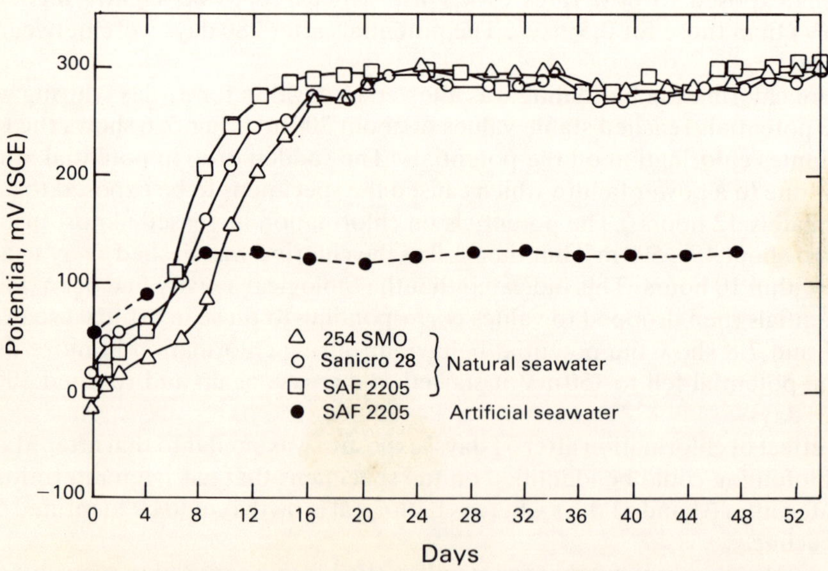

Fig. 7.4 — Comparison of artificial seawater and natural seawater.

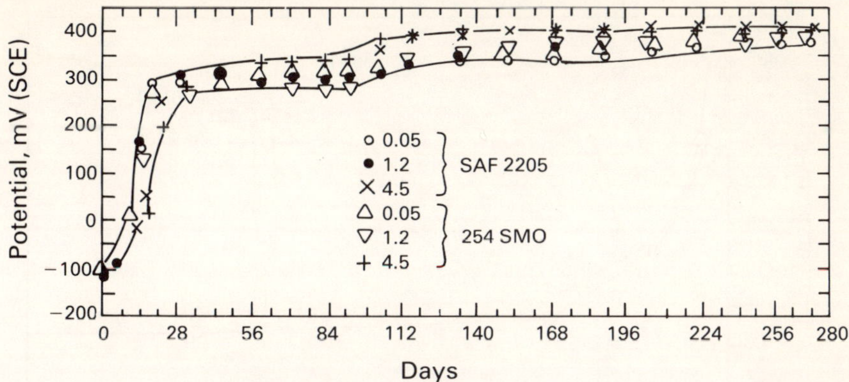

Fig. 7.5 — Corrosion potential for 254SMO and SAF 2205 in seawater at different flowrates.

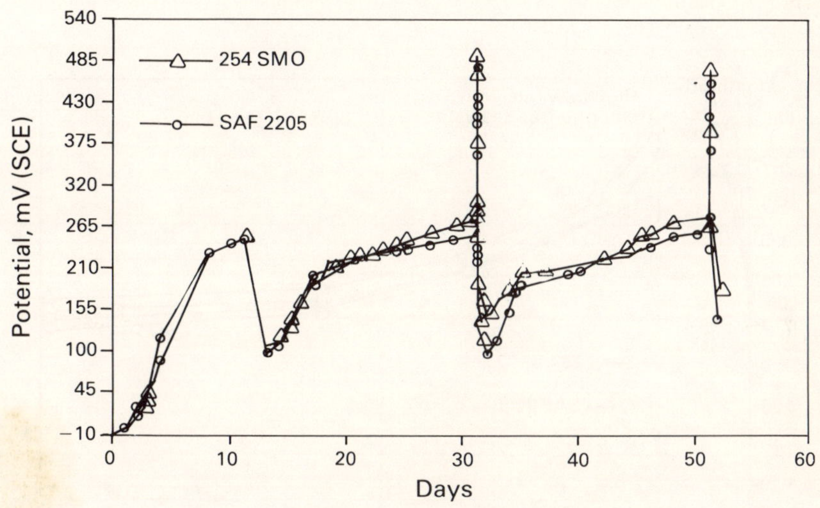

Fig. 7.6 — Potential development in seawater. Chlorination after 31 and 51 days.

2. 350 mV (biofouling);
3. 500 mV (chlorination).

The potential value of 500 mV during chlorination should not be taken as an exact value. According to Wallen [7] cases have been reported where the potential has reached values above 600 mV. High values have also been reported in a literature survey [2].

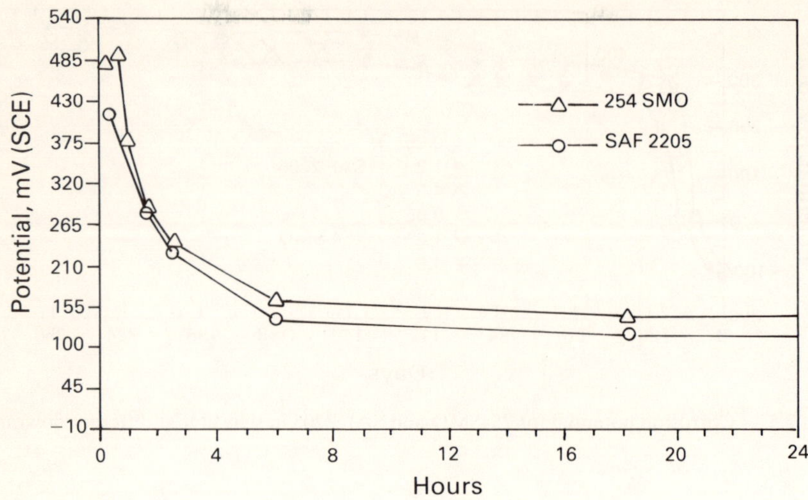

Fig. 7.7 — Same as Fig. 7.6 after chlorination, day 31.

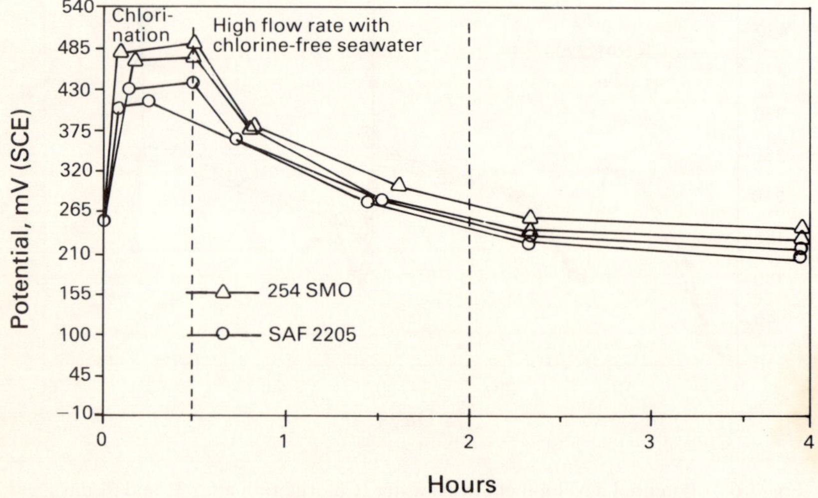

Fig. 7.8 — Same as Fig. 7.7.

Fig. 7.9 [8] illustrates the effect of chlorine content on the corrosion potential of 254SMO in artificial seawater. The rate at which the potential increases is very much dependent on the chlorine content but when the potential reaches a stable value there is no significant difference.

The CPT and CCT for Sanicro 28 and Type 316 polarized from 100 to 600 mV in deaerated 3 per cent NaCl are illustrated in Figs 7.10 and 7.11 [9].

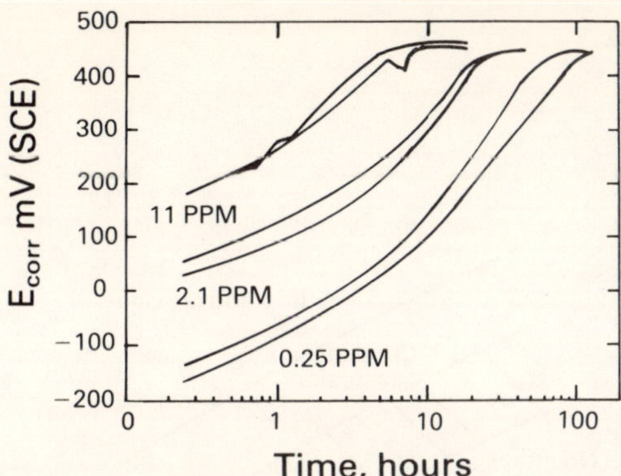

Fig. 7.9 — Effect of chlorination on the corrosion potential of 254SMO in artificial seawater at 25°C. Duplicate specimens.

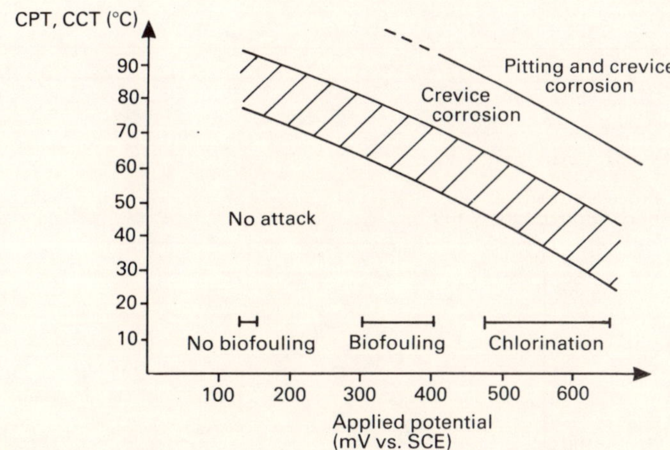

Fig. 7.10 — Critical pitting and crevice temperature for Sanicro 28 at different applied potentials in deaerated 3 per cent NaCl.

It is still too early to say whether the critical temperatures indicated in the figures are relevant to practical applications. However, the data clearly indicate the corrosive effects of both biofouling and, especially, chlorination. Thus, for example, the critical temperature for localized corrosion is lowered by about 40°C in chlorinated seawater compared with seawater with no biological activity.

Sanocro 28 is today performing well in several applications where seawater is used as a cooling medium [10]. Fig. 7.12 illustrates the practical experience with Sanicro 28 in seawater plotted as a function of process stream and seawater

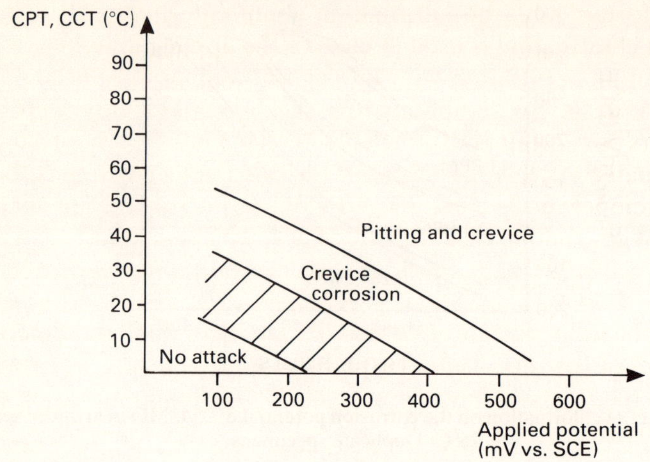

Fig. 7.11 — Critical pitting and crevice temperature for Type 316 at different applied potentials in deaerated 3 per cent NaCl.

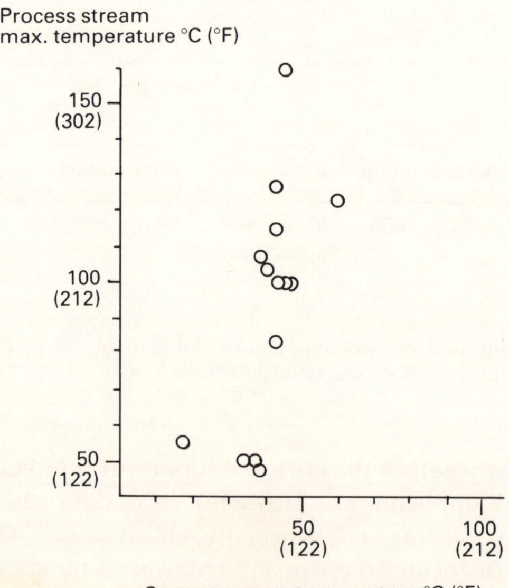

Fig. 7.12 — Practical experience with Sanicro 28 in seawater, plotted as a function of process stream and seawater temperature.

temperature. Unfortunately, the information about chlorination is meagre, but a good guess is that chlorination is used in most of the applications.

Some details of an application with the highest temperature used so far are as follows. In early 1982 Sanicro 28 tube and tube sheet (welded joints) were installed in compressor coolers on a gaslift platform in the Persian Gulf. Seawater at a velocity of 2–$2.5\,\mathrm{ms^{-1}}$ chlorinated at a level of 0.5 ppm Cl_2 was in contact with the tube side. The inlet and outlet temperatures were 30 and 45°C respectively. The shell side was exposed to natural gas containing 2 per cent H_2S and 5 per cent CO_2 at a pressure of 50 bar (725 psi; $5.0\,\mathrm{MPa}$) with inlet and outlet temperatures of 160 and 54°C respectively.

The above application indicates that the CPT curve for Sanicro 28 is conservative as the temperature of the tube wall at the inlet is approximately 95°C, whereas the CPT according to the electrochemical test is 70–80°C at potentials corresponding to chlorination conditions.

EXPOSURE OF HIGHLY ALLOYED STAINLESS STEELS IN CONTINUOUSLY CHLORINATED SEAWATER

Experimental

All specimens were exposed to North Sea water at Bergen at 35°C. The water was continuously chlorinated to give a residual chlorine content of $2\,\mathrm{mg\ Cl_2 l^{-1}}$ (see Fig. 7.2).

The following types of specimens were exposed.

1. Tube-to-tube plate joints of highly alloyed stainless steels. The various combinations of stainless steel tubes and tube plates and methods for joining them together are listed in Table 7.4. The tubes of 254SMO and Monit (24 mm o.d., 0.7/0.8 mm wall) were longitudinally welded and those of Sanicro were seamless.

Table 7.4 — Combinations of stainless steel tubes and plates and methods of joining

Tube plate	Tube material and joint type	254SMO				Sanicro 28				Monit				
		1	2	3	4	1	2	3	4	1	2	3	4	5
35 mm Type 316			×	×		×	×				×	×	×[a]	
35 mm Type 904L		×	×	×		×				×	×	×		
35 mm 254SMO		×	×	×	×[a]									
35 mm Sanicro 28						×	×	×	×					
1 mm Monit											×	×	×	×

1 Roller expansion to 10–14%.
2 Roller expansion to 2–4% followed by perfect TIG-welding without filler.
3 Similar to type 2 but with a simulated weld defect (pipe).
4 Similar to type 2 but TIG-welded with filler normally matching the parent metal (exceptions are given in table).
5 Similar to type 2 but with a simulated repair weld.
[a] Austenitic P 12 filler (type Alloy 625).

2. Creviced samples intended to simulate flange joints with rubber asbestos gaskets. The specimens used included Type 904L, SAF2205, 254SMO, Sanicro 28 and Monit as well as the copper-based material Cu/Ni 90/10. Rubber asbestos washers (dia. 25 mm) were applied to 100×150 mm plates at a pressure of 50 MPa (ASME 8, Division 1, App 2, 1974).
3. Galvanic couples. Combinations of Cu/Ni 90/10 and NiAl bronze coupled to Type 316 and 254 SMO and Ti/254SMO with various area ratios were exposed.

The results from these tests are summarized in Tables 7.5–7.9.

Table 7.5 — Corrosion attack in tube to tube plate joints with 254SMO tubes

Tube plate material	Joint type according to Table 7.4							
	1		2		3		4	
	A	B	A	B	A	B	A	B
Type 316			0/2	0/6	0/2	0/6		
904L	2/2	2/2	1/2	0/6	2/2	0/6		
mm in weld	(1)	(2)	1.5		1.6			
mm in HAZ			0		0			
254SMO	2/2	3/6	0/2	0/6	0/2	0/6	0/2	0/6

(1) Crevice attack <0.5 mm in tube and tube plate.
(2) Crevice attack <0.01 mm in tube and tube plate.

Table 7.6 — Corrosion attack in tube to tube plate joints with Sanicro 28 tubes

Tube plate material	Joint type according to Table 7.4							
	1		2		3		4	
	A	B	A	B	A	B	A	B
Type 316			0/2	0/6	0/2	0/6		
Sanicro 28	2/2	3/6	0/2	0/6	0/2	0/6	0/2	1/6
mm in weld	(1)	(2)						0.4
mm in HAZ								0
904L	2/2	3/6						
	(1)	(2)						

(1) Crevice attack <0.5 mm in tube and tube plate.
(2) Crevice attack <0.01 mm in tube and tube plate.

Table 7.7 — Corrosion attack in tube to tube plate joints with Monit tubes

Tube sheet material	Joint type according to Table 7.4									
	1		2		3		4		5	
	A	B	A	B	A	B	A	B	A	B
Type 316			2/2	6/6	2/2	6/6	0/2	0/6		
mm in weld			1.5	1.0	1.5	1.0	(1)	(1)		
mm in HAZ			0	0	0					
904L	2/2	6/6	2/2	4/6	2/2	5/6				
	(2)	(3)								
mm in weld			1.0	0.1	1.0	0.1				
mm in HAZ			0	0	0	0				
Monit			2/2	3/6	1/2	3/6	2/2	2/6	1/2	2/6
mm in weld			0	0	0	0	0	0	0	0
mm in HAZ			0.1	0.1	0.1	0.1	0.1	0.1	0.1	0.1

(1) P 12 filler (type Alloy 625).
(2) Crevice attack <0.5 mm in tube and tube plate.
(3) Crevice attack <0.01 mm in tube and tube plate.

Table 7.8 — Crevice corrosion attack in chlorinated seawater ($2\,\mathrm{mg\,l^{-1}}\,Cl_2$) at 35°C, 3 months, and critical crevice corrosion temperature (CCT) in 10% $FeCl_3.6H_2O$, 24 hours

Material	Number of crevices attacked	Maximum depth (mm)	CCT (°C)
904L	4/4	0.5	12.5
SS2377	3/4	0.7	17.5
Sanicro 28	3/4	0.5	20.0
254SMO			
sheet	0/4	0	42.5
casting	0/4	0	not tested
forging	0/4	0	not tested
HIP steel	1/4	0.18	not tested
Monit	0/4	0	42.5
90/10 Cu/Ni	0/4	0	not tested

Table 7.9 — Results of tests with various bimetallic couples

Material couples and area ratios	Corrosion rate (μm/year)	Maximum depth of galvanic attack
90/10 Cu/Ni : 90/10 Cu/Ni, 1:1	23.2/24.3	—
90/10 Cu/Ni : uncoupled	19.7	—
90/10 Cu/Ni : 254SMO, 1:1	35.2/0	0.3/0
90/10 Cu/Ni : 254SMO, 1:6	138.1/0	1.0/0
90/10 Cu/Ni : Type 316, 1:6	121.7/0	1.0/0
NiAl-bronze, as cast, uncoupled	5.8	—
NiAl-bronze, heat treated,[a], uncoupled	3.2	—
NiAl-bronze, as cast : 254SMO, 1:6	135.5/0	0.7/0
NiAl-bronze, heat treated[a], : 254SMO, 1:6	104.8/0	0.8/0
NiAl-bronze, heat treated[a] : Type 316, 1:6	121.4/0	0.7/0
254SMO : titanium, 1:6	not determined	0.2/0

[a] Heat treated 2.5 hours at 675°C, air cooled.

In the vase of specimens of type A the results are given in Tables 7.5–7.7. In these the corrosion is recorded as the number of specimens attacked out of the total number of exposed joints, e.g. 2/6, and also as the maximum depth of localized corrosion (mm) in the weld or the heat-affected zone.

Column A in these tables refers to testing according to ASTM G 48–76 and column B to exposure in chlorinated seawater.

For specimens of type B — the creviced samples — the results are given in Table 7.8. For comparison this table includes critical crevice corrosion temperatures obtained in 10 per cent $FeCl_36H_2O$ after 24 hours.

The results for the galvanic couples — type C specimens — are shown in Table 7.9.

This test has been reported in detail by S. Hendriksson [11].

CONCLUSIONS

Chlorine enhances the corrosivity of seawater towards stainless steel. It is therefore important to recognize the oxidizing effect of chlorine when it is applied to installations where stainless steels are used. For such installations the use of intermittent chlorination with as low a dosage as possible is recommended. The time during which the intermittent chlorination is applied should also be minimized since there are indications that the increase of potential is not instantaneous (Fig. 7.9).

Intermittent chlorination applied as described should not raise the corrosion potential above 500 mV at any moment, whereas continuous chlorination will probably lead to a constant potential above 600 mV, which is certainly undesirable for stainless steels.

The effect of chlorination is not only to be considered as negative. For crevice-free construction, e.g. tube heat exchangers with welded tube-to-tube plate joints, the chlorine impedes the formation of crevices that would be caused by biofouling and thereby increases the critical temperature for localized corrosion, i.e. by eliminating the risk of crevice corrosion.

REFERENCES

[1] Dahl, L. Corrosion in chlorinated seawater, Thermal Engineering Research Association, Box 6405, S-113 82, Stockholm, Sweden. *Materialteknik 137*, April (in Swedish) 1983.

[2] Dahl, L. ibid. 209 (in Swedish), 1985.

[3] Bernhardsson, S., Mellstrom, R. and Brox, B. Corrosion 80, NACE meeting, Preprint No. 85. NACE, Houston, Texas, USA, 1980.

[4] Johnsen, R. Slime formation on stainless steel in seawater — the effect of depth and access of light, *SINTEF report STF 16 A85060*, June 1985 (in Norwegian). SINTEF N-7034 Trondheim, Norway.

[5] Garland, P. O. Stainless steel in seawater. The effect of temperature and flow rate on the formation of a slime layer and on the electrochemical properties. *SINTEF preliminary report*, Dec. 1985 (in Norwegian).

[6] Johansen, B. and Supphellen, A. The effect of chlorination on slime formation and the electrochemical properties of stainless steel. *SINTEF report STF34 F85069*, July 1985 (in Norwegian).

[7] Wallen, B. (Avesta), private communication.

[8] Wallen, B. and Abrahamsen, T. Avesta 254SMO. A stainless steel for North Sea applications, *Nordic Symposium on Corrosion in Seawater systems*, IFE, Norway, 27–28 Aug. 1985.

[9] Internal investigation, AB Sandvik Steel.

[10] Sandvik Sanicro 28 seamless tubes. References and booked orders for seawater systems in the process industry, April 1985.

[11] Hendriksson, S. *Highly alloyed stainless steel in chlorinated seawater*, Thermal Engineering Research Association (SVF), 1988.

Part III
Applications

8

Corrosion and chlorination in materials for offshore seawater systems†

R. E. Malpas, P. Gallagher and E. B. Shone
Shell Research Ltd, Thornton Research Centre, PO Box 1, Chester CH1 3SH

INTRODUCTION

The seawater handling systems on offshore platforms serve a variety of tasks: they supply cooling water to such equipment as crude oil and gas coolers, service water for drilling operations and downhole injection, as well as supplying water for fire mains systems and living quarters. To control fouling such systems are routinely chlorinated. In this chapter we present a description of chlorination practice in existing offshore seawater handling systems which are, in the main, constructed from copper–nickel alloys. This is followed by an account of some aspects of our research into the use of alternative materials for fabrication of these systems (stainless steels) and the effect that chlorination could have on the corrosion characteristics of these alloys.

EXISTING SYSTEMS

The seawater handling systems used on most offshore platforms in the North Sea are in the main constructed from copper alloys and have reached a state of development where in general their performance is regarded as being very satisfactory. The only causes of failures are those associated with either excessive seawater velocities and turbulence (giving erosion and leakage) or fatigue. To avoid these problems seawater velocities have to be maintained below $3.5\ \text{ms}^{-1}$ while thick-walled pipe is used to prevent buckling of those alloys which have low tensile strengths and moduli of elasticity.

Failures are inevitably found to be related to non-compliance with these guide-

† This chapter is published under licence granted by Shell Research Limited, Thornton Research Centre, PO Box 1, Chester CH1 3SH.

lines during design. To reach this stage of development a considerable amount of research has been carried out over many years to establish the limitations of copper alloys. Additionally these alloys have been used in marine applications for many years and information relating to most aspects of their performance in aerated seawater is well documented [1,2].

Such seawater systems are chlorinated to prevent fouling which, if not checked, can lead to blockage (which can occur in copper– nickel systems after as little as 12 months' service). In critical areas such as heat exchangers fouling can induce localized attack of the type shown in Plate 8.1 where severe pitting of a 70:30-type

Plate 8.1 — Barnacle damage on condenser tubes.

copper–nickel condenser tube was observed under an attached barnacle. This was presumably caused by differential aeration effects and would in a very short time lead to tube perforation.

Chlorination is most commonly carried out using electrochlorination units. Operating experience of these units has been gained over the past 15 years in vessels in the Shell fleets where various problems became apparent, including platinum loss from electrodes, flow switch malfunctions, cell and pipework leakage, cell blockage by calcareous deposits and complicated interactions of each of these. These have been described in a recent publication [3]. For shipboard application feedwater chlorine concentrations of 0.5 ppm, resulting in a residual chlorine level of 0.1 ppm at   main condenser outlets, are generally sufficient with either periodic (deep sea) or continuous (harbour) dosage.

On offshore structures, however, a somewhat higher residual chlorine level of 0.5–1.0 ppm is maintained to provide a greater error margin in these larger systems, where considerable quantities of water may be stored at a time. Chlorine is introduced into the caisson below each submersible feed pump with suitable fail-safe valves to prevent a potentially harmful build-up of chlorine in the event of pump failure. In general, provided chlorine levels are maintained in the range stated above, chlorine-related corrosion problems are not encountered.

FUTURE SYSTEMS

As oil and gas production extends into deeper waters and less expensive reserves are produced, it will become increasingly more and more important to reduce weight on the topside of platforms. It has been estimated, for example, that reduction of 1 tonne of weight above the water line could result in savings of around £100 000 on the cost of steelwork in the subsea jacket [4]. One area in which weight saving seems possible is that of the thick-walled, and thus heavy, copper alloy seawater handling system.

From manufacturers' claims the 'newer' high alloy stainless steels appear to be attractive candidate materials for fabrication of seawater systems. Information relating to the performance of these materials in real seawater environments, however, is either somewhat limited or not readily available from independent sources. However, the work of Hack, Streicher and others [5–9] indicates that these materials certainly must be considered for seawater service applications.

A few years ago we began a research programme which was aimed at establishing the possibility of using stainless steel alloys for the construction of compact, lightweight seawater handling systems. Our objective is to design a system from compatible components as opposed to one built from collections of what may be well-engineered components that are incompatible from a corrosion viewpoint. With this in mind the testing of pumps, valves, heat exchangers and other small components, as well as pipework, has been included in our research programme. The need for chlorination in these systems has meant that the influence of differing levels of chlorine on the corrosion properties of these material has been one area for study and has to date produced some interesting results.

Initial screening tests for resistance to crevice corrosion were carried out in our dockyard test facility on over 40 different alloys (some of which are listed in Table 8.1) in flowing natural (unchlorinated) seawater using a multiple crevice washer test.

Corrosion potentials were monitored throughout the 100-day crevice washer tests and the area and depth of crevice corrosion determined afterwards. The results obtained with the different alloys can be broadly classed into three groups.

1.   Those materials where the corrosion potential rose initially over the first 7–20 days to about 300 to 350 mV (SCE) and then remained effectively constant for the remainder of the test period. Fig. 8.1(a) shows a plot of potential against time for 254SMO as an example of this group. Other alloys in this group included Zeron 100,

Table 8.1 — Trade names and chemical analyses of some of the alloys tested

Alloy trade name	Manufacturer	Category	Chemical composition, % wt										
			Cr	Ni	Mo	N	Mn	C	Si	P	S	Cu	Others
316L	BSC	Austenitic	17.1	11.6	2.06	0.03	1.16	0.028	0.34	0.03	0.003	—	—
317LN	BSC	Austenitic	18.8	13.5	3.82	0.15	1.64	0.015	0.57	0.027	0.01	—	—
904L	BSC	Austenitic	19.2	25.4	4.2	0.03	1.59	0.01	0.24	0.015	0.002	1.57	—
254 SMO	Avesta	High alloy austenitic	19.3	18.3	5.94	0.215	0.50	0.017	0.56	0.025	0.003	0.70 Co 0.2	Nb 0.04 Ti>0.1
Cronifer 1925 hMo	VDM	High alloy austenitic	20.65	24.75	6.24	0.134	1.43	0.014	0.29	0.022	0.003	0.80	—
Cronifer 2205 LCN	VDM	Duplex	22.3	5.5	2.77	0.118	1.73	0.01	0.31	0.023	0.006	—	—
SAF 2205	Sandvik	Duplex	22.5	5.59	2.98	0.15	1.65	0.025	0.35	0.024	0.003	—	—
WI.4462	Nyby Uddeholm	Duplex	21.9	5.15	2.77	0.114	1.19	0.025	0.49	0.026	0.011	0.16	—
Ferralium 255	Cabot	Duplex	25.5	5.1	3.03	0.17	0.65	0.020	0.35	0.017	0.006	1.75	—
DP3	Sumitomo	Duplex	25.1	7.2	3.14	0.152	0.89	0.021	0.41	0.027	0.002	0.51	W 0.28 Ti 0.46
Monit	Nyby Uddeholm	Superferritic	25.5	4.0	4.02	0.023	0.25	0.014	0.29	0.02	0.003	—	—
Zeron 100	Mather and Platt	Duplex	24.4	6.7	4.0	0.22	0.38	0.036	0.31	0.019	<0.003	0.55	W 0.8

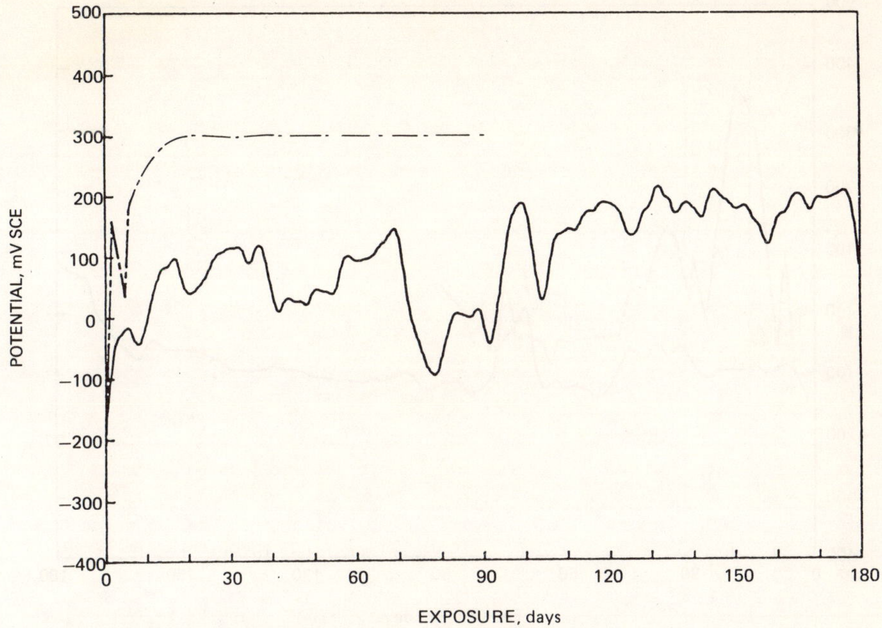

Fig. 8.1 — Potential/time relationships for 254 SMO alloy: (a) (− · −) crevice washer tests in unchlorinated seawater; (b) (——) system test in chlorinated (0.1 ppm) seawater.

Ferralium 255, Cronifer 1925 hMo, and DP3. All these materials showed no, or only a minimal amount of, crevice corrosion after 100 days of exposure. This is shown schematically in Fig. 8.3.

2.   Those materials where corrosion potentials rose initially to a potential in the range 100–300 mV SCE but then fell sharply to around 0 mV. The potential then remained in the 0–200 mV region with oscillations. Fig. 8.2(a) shows a potential-time plot for the 22Cri-5Ni duplex stainless steel W1.4462 as an example of this group. Most of other low chromium duplex stainless steels showed this behaviour characteristic. These materials showed a slight amount of crevice corrosion after 100 days of exposure (see Fig. 8.3).

3.   Those materials where potentials remained low (between 0 and −100 mV SCE) throughout the duration of the test. Fig. 8.4(a) shows a potential-time plot for an AISI 316L alloy as an example of this group. Most conventional austenitic steels showed this type of behaviour. All these materials showed extensive crevice corrosion after 100 days of exposure (see Fig. 8.3).

From these initial tests a number of alloys representative of the three groups (which we designate group 1, 2, or 3 materials) have been chosen for extended testing in prototype seawater systems. The design of the initial system chosen is shown in Fig. 8.5. As many typical features of seawater piping systems as possible have been incorporated and, since it was intended to test the materials under the most adverse

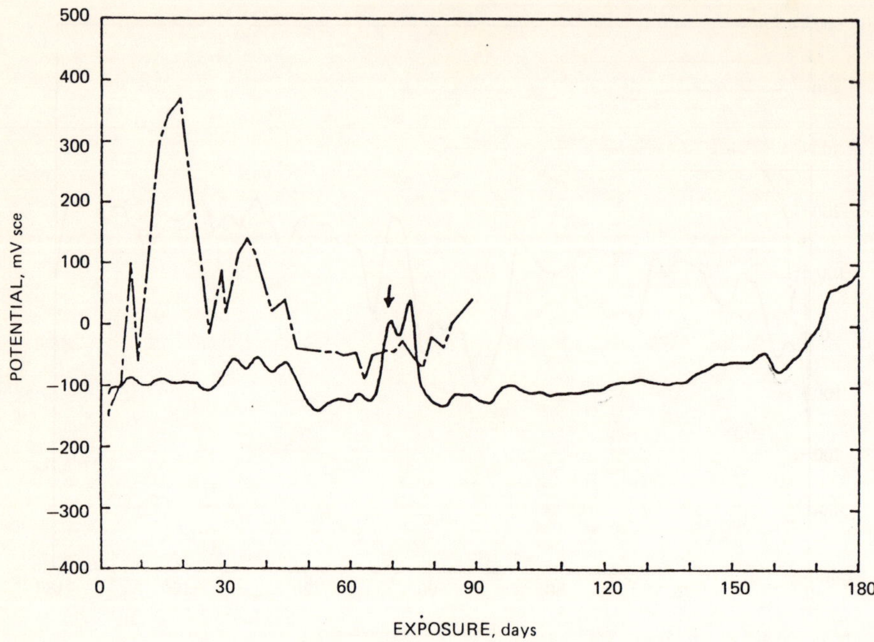

Fig. 8.2 — Potential/time relationships for 22Cr-5Ni duplex alloy: (a) (– · –) crevice washer test on W1.4462 in unchlorinated seawater; (b) (——) system test on SAF 2205 in chlorinated (0.1 ppm) seawater. Failure of the electrochlorinator occurred after 67 days (arrowed).

conditions, a number of examples of bad engineering practice were also included, such as pipe branching, deadlegs and faulty welding. Both wrought and cast flanges were used where possible, with standard fibre gaskets; while welds were made using either gas tungsten arc welding (GTAW) or shielded metal arc welding (SMAW) with either matching or more noble filler metal.

Seawater was pumped through these systems at a velocity of 0.1 m/s using one of two pumps that were themselves constructed from two of the materials under test. The seawater in these tests was chlorinated, to an inlet total chlorine level of 0.5–0.8 ppm. This resulted in a residual chlorine level of around 0.1 ppm. The tests were carried out for a period of 6 months during which time the potential of the system was monitored continuously.

The potentials recorded during the tests in systems constructed from 254SMO, a 22Cr-5Ni duplex material (SAF2205) and AISI 316L alloys (i.e. the representative materials of the above three groups) in this chlorinated seawater, are shown in Figs 8.1(b), 8.2(b), 8.4(b). As can be seen, the potentials observed in these tests were considerably lower than those seen in the unchlorinated crevice washer tests, although the values observed were still in the order:

254SMO > SAF2205 > AISI 316L

That this drop in potential is caused by chlorination is demonstrated in Fig. 8.2(b)

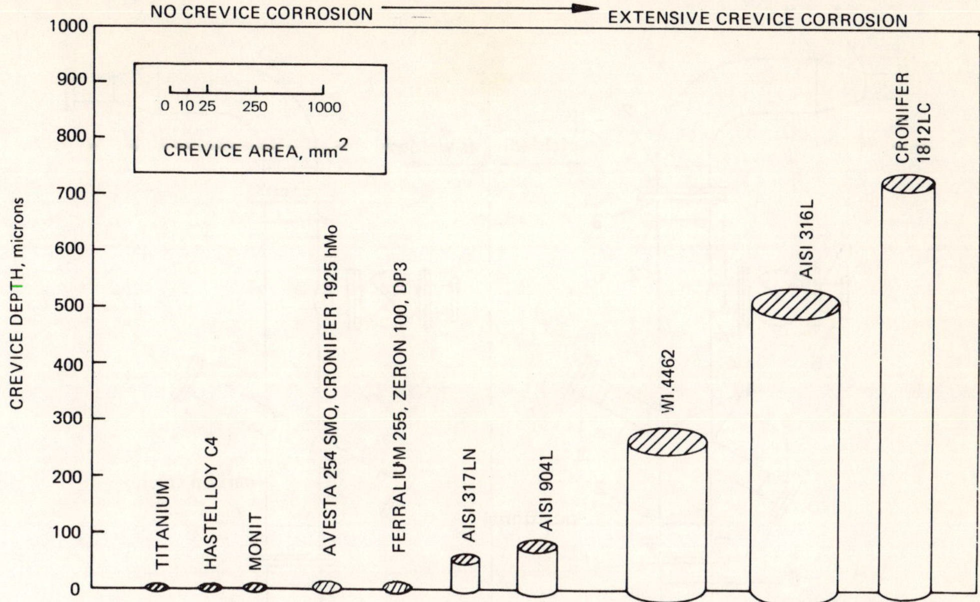

Fig. 8.3 — Crevice depth and area of attack on alloys after 100 days exposure in unchlorinated seawater. The area of attack is indicated by the cylinder width (see inset scale).

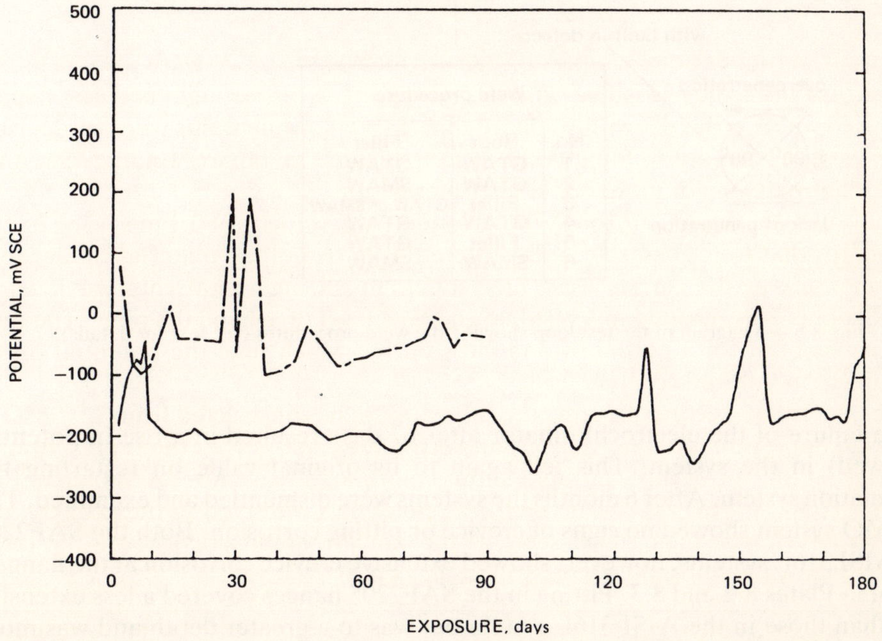

Fig. 8.4 — Potential/time relationships for AISI 316L alloy: (a) (—·—) crevice washer test in unchlorinated seawater. (b) (———) system test in chlorinated (0.1 ppm) seawater.

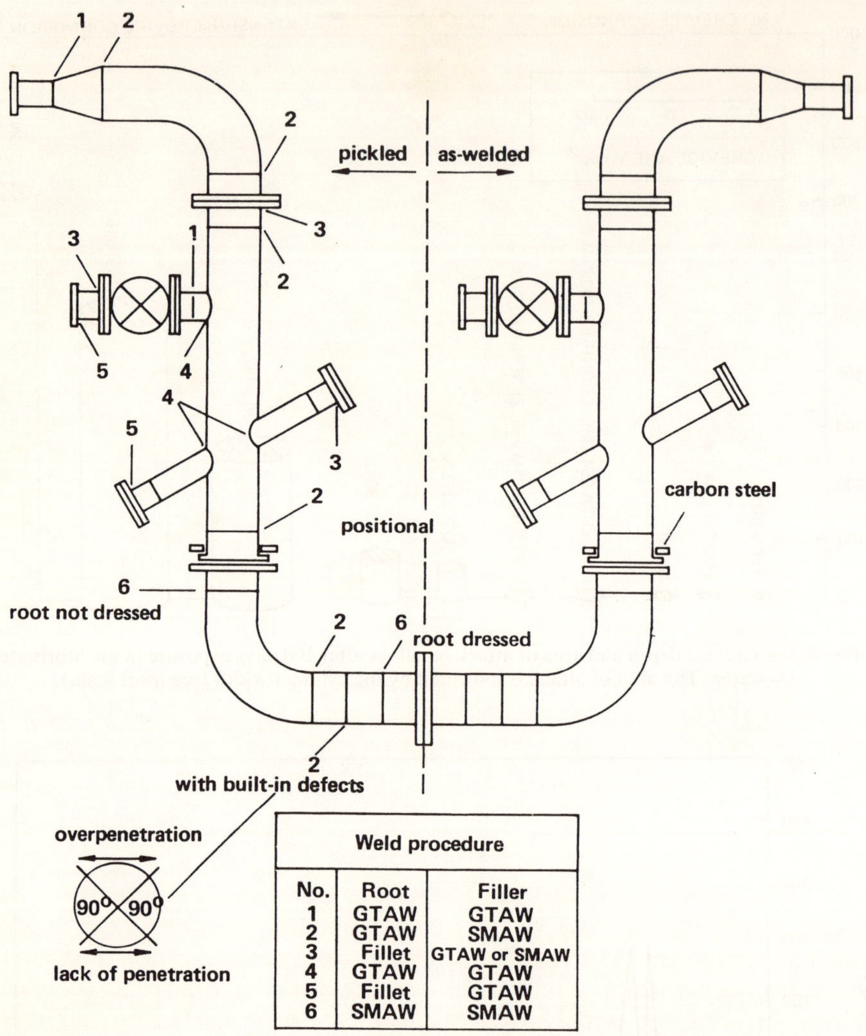

Weld procedure		
No.	Root	Filler
1	GTAW	GTAW
2	GTAW	SMAW
3	Fillet	GTAW or SMAW
4	GTAW	GTAW
5	Fillet	GTAW
6	SMAW	SMAW

Fig. 8.5 — Diagram of the test loop showing the weld procedures (see text for details).

where failure of the electrochlorinator after 67 days resulted in a rise in potential (arrowed) in the system. This fell again to its original value on restarting the chlorination system. After 6 months the systems were dismantled and examined. The 254SMO system showed no signs of crevice or pitting corrosion. Both the SAF2205 and AISI 316L systems, however, showed extensive crevice corrosion at the flanges, shown in Plates 8.2 and 8.3. Pitting in the SAF2205 flanges covered a less extensive area than those in the AISI 316L system but was to a greater depth and was more severe on the cast flanges. Some isolated pitting of the AISI 316L pipework was also observed; however, on all systems no corrosion was observed at welds (despite the deliberate defects) or in the vicinity of any of the deliberate design faults.

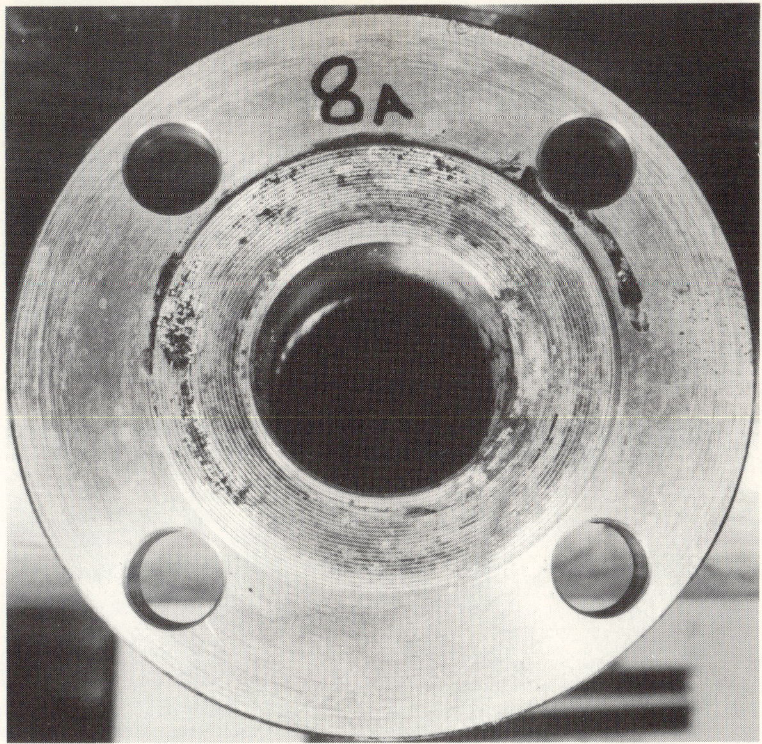

Plate 8.2 — 316L pitted flange.

Our tests in larger systems appear, therefore, to have confirmed the superior seawater corrosion resistance of 'group 1' materials over the 'group 2' 22Cr5Ni duplex steels or 'group 3' conventional austenitic alloys. The alloy 254SMO has stood up to our testing particularly well. The lower potentials observed in these chlorinated systems compared to those recorded in the unchlorinated crevice washer tests are very interesting, however, and, we believe, could have important implications for the corrosion behaviour of, in particular, the 'group 2' alloys in chlorinated seawater.

Considering first the potentials observed from alloys in the unchlorinated tests, we suggest that the stable potentials obtained with 'group 1' materials in the region of 300–350 mV SCE represent the equilibrium rest potentials of these materials and that, in the absence of the initiation of any localized corrosion, all the materials studied here (which are expected to have chemically similar surfaces) would show rest potentials in this region. This potential, and the time taken to attain it, will depend critically on experimental conditions and, in particular, on those factors influencing the cathodic (oxygen reduction) reaction kinetics. Temperature, for example, would be expected to have an effect, and in fact in winter we routinely observe lower rest potentials and longer times for these to be attained.

Attainment of a stable rest potential by an alloy, however, will depend on whether localized corrosion occurs or not, as the presence of a site of active corrosion

Plate 8.3 — 2205 pitted flange.

will strongly influence the potential of the system, shifting it to more negative values. In turn, the occurrence of localized corrosion at any potential depends on the probability of initiation of localized corrosion (i.e. the pitting probability) of the material at that potential.

By this argument, therefore, 'group 1' materials are those with a low pitting probability at high potentials, the equilibrium rest potential of the material can thus be safely reached and maintained with only a very small chance of pit initiation. Conversely 'group 3' materials, such as AISI 316L, have a high pitting probability at low potentials.

Pitting thus initiates easily and prevents attainment of the equilibrium rest potential. Between these two extremes come the 'group 2' materials such as the 22Cr5Ni duplex steels. These alloys can, in many cases, reach their rest potentials; however, their pitting probabilities at these potentials are high, resulting in the initiation of pits and subsequent lowering of the potential. Any repassivation of the pits in these materials will, of course, allow the potential to rise again before a fresh pit initiates. Oscillation of the potential is thus observed in these cases.

In chlorinated systems with low residual chlorine levels the equilibrium rest potentials of these alloys (represented by the stable potentials of 'group 1' materials) are lowered to around 100–200 mV SCE.

In general we would expect pitting probabilities to reduce as the potential is lowered; the lowering of the potential of these materials on chlorination might therefore be expected to lead to less localized corrosion. In practice, however, this observation will probably depend on the type of alloy. With 'group 3' alloys, pitting probabilities appear to be high even at this lower potential and little change is expected in corrosion behaviour on chlorination. Similarly, little change is expected in the performance of 'group 1' alloys which are already in a regime of low pitting probability in this potential region (although their resistance to pitting may be strengthened by this shift to lower potentials). 'Group 2' materials, however, which show high pitting probabilities at high potentials and a significantly lower pitting probability as the potential is reduced, should be most influenced by this shift in potential on chlorination. Thus, provided the potential is lowered to a significant extent it is possible that initiation of pits in these materials may be considerably delayed or even prevented from occurring. Chlorination to these low residual levels (0.1 ppm) should thus be beneficial to the corrosion performance of these 'group 2' alloys.

The reason for the drop in potential on chlorination is uncertain; however, one likely explanation is a change (a decrease) in the kinetics of the cathodic (oxygen reduction) reaction accompanying the death of marine organisms on chlorination. It has been suggested previously [10] that these organisms can catalyse oxygen reduction, possibly by enzyme action, resulting in a relatively positive mixed rest potential in the system. It is interesting to note that the potentials observed in these chlorinated systems are very similar to those seen in artificial seawater where marine organisms are also absent. We have observed many times in the past that corrosion in artificial seawater is considerably less than that seen in natural seawater, lending support to our ideas about the advantage of chlorination.

Finally it should be recognized that these possible advantages of chlorination will only apply at low residual levels. As chlorine concentrations rise above 0.1 ppm, chlorine reduction depolarizes the reaction and the potential of the system will rise. At very high chlorine levels the extent of corrosion could be severe. The general potential–chlorine relationship resembles Fig. 8.6.

To minimize corrosion, while maintaining adequate anti-fouling properties in stainless steel seawater systems, it would thus appear that chlorination levels should be kept in the region of this potential minimum which we believe to be around the 0.1 ppm residual level although control of chlorination to this level in large systems may be difficult. Further work is currently in progress to quantify this relationship on our complete range of alloys as well as testing the materials at these low chlorination levels in larger prototype seawater systems at high flow rates.

REFERENCES

[1] Shone, E. B., Problems in seawater circulating systems, *Br. Corros. J.*, (1), 32, 1974.

[2] Connell, R. A. and Shone, E. B., Seawater circulating systems, Supplement to *Chemistry and Industry*, 2 July 1977.

[3] Shone, E. B. and Grimm, G. C., 25 years experience with seawater cooled heat transfer equipment in the Shell Fleets, *Trans I. Mar. E.* (TM), **98**, paper II, 1985.

[4] *The North Sea and British Industry: the new opportunities*. A report commissioned by Shell UK Ltd from the Economist Intelligence Unit, April 1984.

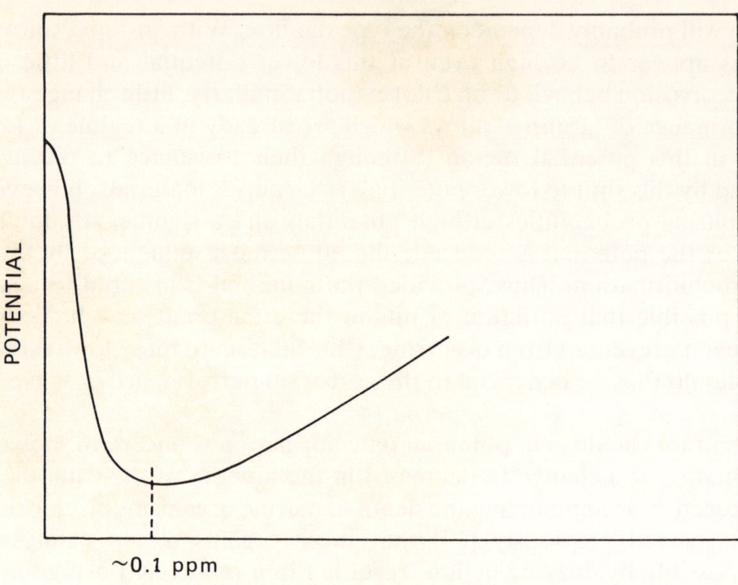

Fig. 8.6 — A scheme representation of the predicted relationship between the potential attained by a stainless steel alloy in seawater in the absence of corrosion and the level of chlorine in the seawater.

[5] Hack, H. P., Crevice corrosion behaviour of 45 molybdenum-containing stainless steels in seawater, *Corrosion* 82, paper 65, NACE, 1982.

[6] Streicher, M. A., Analysis of crevice corrosion data from two seawater exposure tests on stainless steel alloys, *Materials Performance*, **22** (5), 37, 1983.

[7] Asphahani, A. I. *et al.*, Highly alloyed stainless materials for seawater applications. *Corrosion* 80, paper 29, NACE, 1980.

[8] Bond, A. P. and Dundas, H. J., Stainless steels for seawater service, *Stainless Steel* 77, paper 15, Climax Molybdenum Co, 1977.

[9] Sedriks, A. J., Corrosion resistance of austenitic Fe-Cr-Ni-Mo alloys in marine environments, *International Metals Review*, **26** (6), 321, 1982.

[10] Scotto, V., Di Cinitio, R. and Marcenero, G., The influence of marine aerobic microbial film on stainless steel corrosion behaviour, *Corrosion Science*, **25**, 185, 1985.

9

Condenser tube corrosion in CEGB seawater-cooled plant: recent service and research experience†

W. E. Heaton

INTRODUCTION

In recent years, the engineering and scientific problems associated with electrical power generation have provided the corrosion scientist and engineer with a rich field for research, development and innovation. In essence, electricity is generated using one form of the steam cycle; at one end of the cycle, the heat production end, corrosion problems may occur, associated, for example, with oxidation of boiler tubes. It is at the other end of the steam cycle, the heat rejection end, that the condenser and its corrosion problems are encountered.

The condenser acts as a heat exchanger between the steam and the cooling water. The exchange is achieved by passing the cooling water through banks of tubes secured between tubeplates in a steel shell. The cooling water is pumped into cast iron or mild steel water boxes and from these flow through the condenser tubes is maintained at the velocity required for optimum heat transfer conditions. The steam encounters the external surface of the tubes, the heat of evaporation is extracted, the steam condenses to the liquid phase and is returned as boiler feed water for reheating. The most important function of a power station steam condenser is to provide the lowest possible temperature at this, the heat rejection end of the cycle.

The size of turbo-alternators has increased from typical units commissioned in 1952 of 60 MWe to units commissioned in recent years of 660 MWe. This increase in

† This chapter is an expanded version of a paper first presented at a Seminar on Prevention of Condenser Failures — The State of the Art, Electric Power Research Institute, Palo Alto, December 1985, and at UK Corrosion '86, Institution of Corrosion Science and Technology, Birmingham, November 1986.

The author was a Research Officer at the General Electricity Generating Board, Operational Engineering Division, Scientific and Technical Branch (Southern Area), Canal Road, Gravesend, Kent, DA12 2Rs.

unit size has been achieved by rises in steam temperature and pressure conditions. Concomitant with these have been increases in the requirements for purity and freedom from contamination in feedwater and boiler drum water specifications.

The reason for the stringency in these specifications is due to the possibility of boiler tube corrosion increasing with the increases in boiler water throughput — so-called 'on-load' corrosion. Any contaminant present could form a deposit on the inner surfaces of the boiler tubes and set up catastrophic corrosion. When it is considered that for a 660 MWe unit over 2×10^6 kg of steam are evaporated every hour, contaminants at a concentration of only 1 in 10^9 could cause a deposit of 1.75 kg to form after only one year's operation. It is essential, then, that the condenser should provide boiler water to the highest possible standards of purity.

One of the most important features of operational efficiency in a turbo-alternator is that the lowest possible pressure should be achieved when the steam is condensed to the liquid phase within the condenser. This occurs when the cooling water is at the lowest possible temperature. The quantity of cooling water required for a fully operational 2000 MW generating station would be of the order 2×10^8 l/h (4.5×10^7 gallons/h).

The source of such vast quantities of cooling water is provided by the physical geography of the country; the present concern with the quality of the environment is reducing the possibility of using inland sources of fresh water for cooling purposes, and coastal or estuarine sites are being increasingly sought and used. The large heat sink available and the comparative narrow seasonal temperature ranges make seawater a very attractive medium for cooling purposes.

The CEGB has a generating plant capacity of about 52 000 MWe, of which nearly half is located at coastal or saline estuarine sites.

A leakage of seawater into the condensate through a leak passage of 10^{-2} mm diameter will allow sufficient seawater (2 litres) into the boiler charge in 1000 hours to contravene the set standards. Such a leak may occur in two ways; first, leakage of seawater from the waterbox through the joints between the tubeplate and the tube ends. This problem has largely been overcome by improved methods of tube/tubeplate fixing and the development of the double tubeplate. The space between the double tubeplate is either at vacuum or filled with condensate under pressure, so that any leakage is into the cooling water and not into the condensate.

The second source of leakage of seawater into the condensate is through defects in the tube wall. Such defects, if due to faulty manufacture, are normally traced and rectified during precommissioning trials, and leakage through condenser tube walls normally gives concern only after some period of operational usage. Condenser leaks usually occur as a consequence of water side corrosion of the condenser tube material in the form of pitting corrosion or erosion–corrosion or a combination of both. At some locations the presence of suspended solids in the cooling water may also lead to abrasive wear and perforation of condenser tubes.

SERVICE EXPERIENCE WITH CONVENTIONAL MATERIALS

Copper alloys, particularly aluminium brass and the copper–nickel alloys, have been used for many years in condenser tube nest construction in view of their good heat transfer properties, their economic viability and, in general, satisfactory corrosion resistance [1]. These materials rely on the formation of a protective surface film for their corrosion resistance in a given environment. Once this film is damaged, corrosion will proceed unless conditions are favourable for film repair.

Pitting corrosion has been observed to occur as a result of the deposition of mud or silt on the tube surface. Beneath these deposits, oxygen access is restricted and if protective film formation is inhibited, pitting corrosion takes place and may be accelerated by the galvanic effect of being in contact with fully oxygenated silt-free areas, the anodic process of metallic corrosion:

$$M \rightarrow M^{n+} + ne^-$$

being supported by the cathodic process, the reduction of dissolved oxygen:

$$O_2 + 2H_2O + 4e \rightarrow 4OH^-$$

The insoluble corrosion products which result may be in the form of adherent nodules, the most frequently observed species being green copper hydroxy chlorides, particularly paratacamite, $Cu_2(OH)_3Cl$, and the binary copper hydroxy chloride $CuCl_2.3Cu(OH)_2$, often admixed with the silt. When these surface deposits are removed the pits are revealed in the surface, which may contain copper-coloured crystalline deposits of cuprous oxide, Cu_2O. The local disturbances in flow created by these nodules and pits may, in turn, accelerate the loss of material by an erosion–corrosion mechanism leading to tube failure. This form of attack following pitting corrosion under silt films has been observed in both aluminium brass and copper–nickel materials.

All seawater and estuarine-cooled plant operate under syphonic conditions, i.e. the water is subject to a negative pressure gradient in its passage through the condenser. Erosion–corrosion is the major cause of copper alloy condenser tube deterioration at these sites and is most frequently encountered as either inlet end attack or as the result of lodged obstructions in the tube bore [2]. The nature of lodged obstructions varies from the common mussel (*Mytilus edulis*), filamentous weed (in particular, the cast hydroid fronds of *sertularia cupressina*, also known as whiteweed), concrete debris, calcareous deposits from cathodic protection systems, coal, and the hard, brittle particles of corrosion products from corroding mild steel components. In such cases, the localized increase of cooling water velocity, due to the reduction in cross-sectional area, in combination with the reduction in pressure due to the syphonic conditions promotes the localized release of dissolved gases. The bubbles of gas impinge and collapse on the tube walls, causing a rapid breakdown in the protective film normally formed on the tube surface allowing erosion–corrosion to proceed and eventually leading to perforation of the tube wall [3, 4]. Figs 9.1 and 9.2 illustrate this effect.

With the continuing incidence of condenser failures in copper alloy tubed condensers, it has become an established practice for regular non-destructive testing to be carried out to detect incipient condenser tube deterioration. The eddy current technique is used for this and modern developments in data acquisition and data

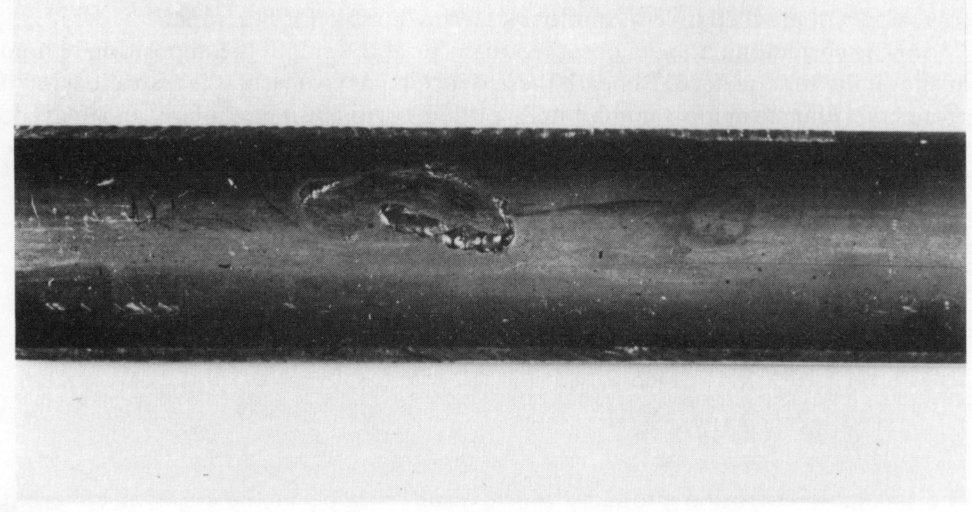

Fig. 9.1 — Erosion–corrosion due to a lodged obstruction.

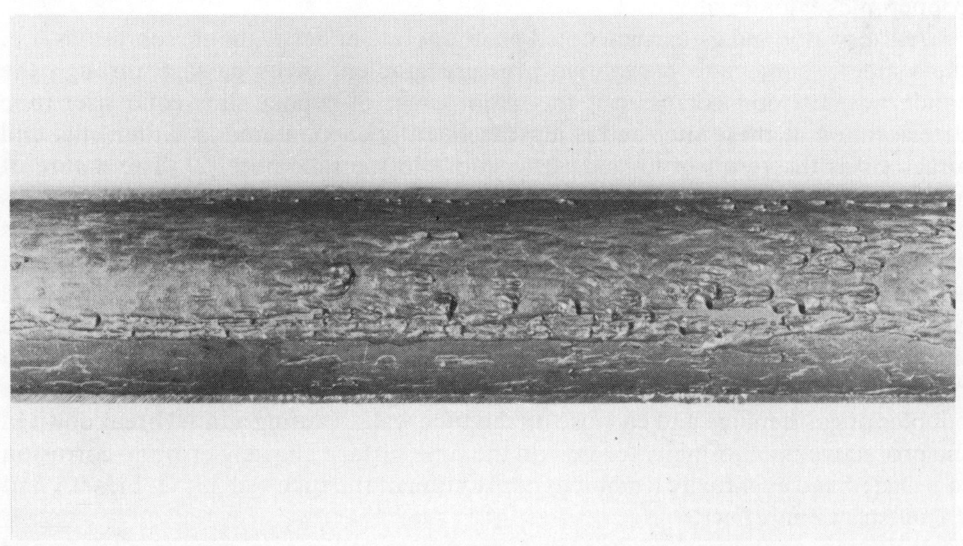

Fig. 9.2 — Erosion–corrosion due to release of dissolved gases under syphonic conditions.

processing have enabled considerable improvements to be made in both the speed of inspection and the presentation of results. It is now possible for parameters such as tube thinning, pit dimensions, pit depths and pit locations to be accurately assessed. Preventative measures such as tube plugging can then be carried out to prevent propagation of such defects taking place which would eventually cause tube perforation and condensate contamination.

When a tube has perforated as the result of pitting corrosion or erosion–corrosion, seawater ingress into the condensate is detected by changes in condensate conductivity. Conductivity monitors are located in the condenser and these give some indication of the location of the leaking tube. The contaminated condensate is passed to the water treatment plant where demineralization takes place. Modern, high efficiency condensate polishing plant can deal with leakages of up to 1 ppm chloride in the condensate but the total volume is limited by the necessity to regenerate the ion exchange resins. The leaking tube must be identified and a variety of off-load measures have been developed for this purpose. The tube then has to be plugged to prevent further condensate contamination. As a result, condenser performance may be adversely affected by the reduction of coolant availability; this usually becomes economically and operationally significant when the number of tubes affected exceeds ~10 per cent.

The economic importance of high-merit, high-output electricity generating plant makes it essential to maintain continuous condenser availability, particularly for nuclear plant where two-shift operation is not feasible. Partial shut-down of passes in the condenser for leak location is sometimes possible, but certain designs do not lend themselves to this operation. In these circumstances, and where condenser corrosion has been an operational problem, it becomes economically viable to consider retubing in materials which, although of higher cost, offer immunity from corrosion problems.

REMEDIAL AND PREVENTATIVE MEASURES

Prevention of debris ingress and mussel infestation is a prerequisite for the maintenance of condenser integrity [5] and hence the effectiveness of the main screening plant is of prime importance (Fig. 9.3). An effective biofouling control regime is also essential, not only to prevent the growth and settlement of mussels (macrofouling) in the cooling water system but also to control biofouling (microfouling) of condenser tube surfaces [6]. Chlorination, whether from gaseous chlorine, sodium hypochlorite or sodium hypochlorite produced by brine or seawater electrolysis, is the most frequently used method and the inhibition of organic biologically active slime films resulting from microfouling on waterside surfaces will also help to prevent the formation of the complex multi-component films associated with the initiation of pitting corrosion.

The combination of the Taprogge (Amertap) debris filter with the sponge ball condenser cleaning system [7] has proved successful as part of a condenser integrity preservation programme. The debris filter consists of a perforated cylindrical stainless steel filter located within a 90° bend in the pipework. Due to the radial flow, debris collects on the outside of the filter basket. In the flushing cycle, the debris is removed by a tangential and reverse radial flow and transferred to discharge

Fig. 9.3 — Typical seawater screening plant.

pipework. The sponge balls for condenser microfouling control are introduced into the filtered seawater and are compressed into and through the condenser tubes by the water flow. The balls are collected in a strainer in the outlet pipework and recirculated in the condenser system.

The promotion of iron-rich protective film formation by ferrous sulphate injection has been used for many years [8,9], but CEGB experience has been that, in many cases, it is ineffective in the long term and there could be additional problems with reduced heat transfer efficiency, although this can be maintained using the Taprogge sponge ball cleaning system.

Metal and plastics inlet end inserts are now well established methods to overcome inlet end erosion–corrosion effects. In recent years, the use of an insert with a venturi design in hygroscopic nylon with a reduced inlet diameter has proved successful in arresting the passage of debris down a tube (Fig. 9.4); waterborne debris small enough to pass through the insert should pass through the tube and should not form a lodged obstruction [10].

A major component of the debris identified as turbulence raisers in condenser

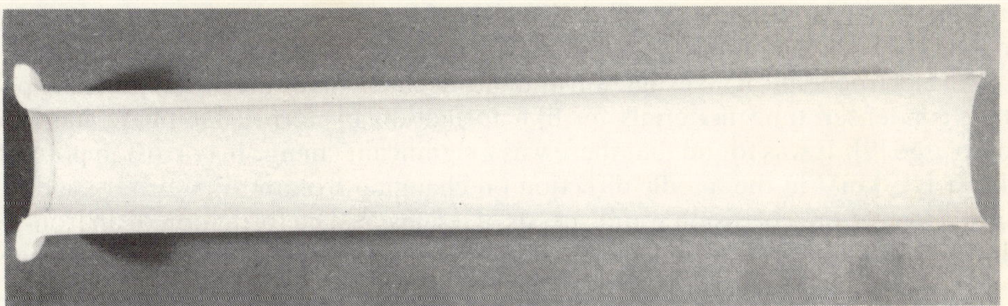

Fig. 9.4 — Cross-section through condenser insert.

tubes are the corrosion products of fabricated mild steel water boxes. In many cases, this has been the consequence of the comparatively poor performance of cathodic protection installations in condenser water boxes as the result of a combination of the shortcomings of the design parameters (in particular, anode disposition), routine surveillance and maintenance of such systems.

The costs involved in the rectification of these defects is so high that it has been a wide practice in recent years for the corrosion control measures for condenser and auxiliary cooler water boxes to be restricted to high grade protective coatings. A multi-coat neoprene preparation has been particularly successful. The combination of protective coatings and impressed current cathodic protection has been avoided because of the possible problems of cathodic disbonding and blistering of the coating, although zinc sacrificial anodes have been used successfully in combination with certain mineral-filled epoxy coatings.

CONDENSER TUBE CORROSION RESEARCH

Recent CEGB research work has been concerned with investigations into corrosion control of conventional materials and the development of alternative condenser tube materials. This work was carried out at the Seawater Corrosion Laboratory which was located at Brighton Power Station. The facilities at this Laboratory enabled corrosion trials and experimental work to be carried out in flowing, once-through seawater. This ensured realistic environmental conditions and precluded errors introduced by artificial media and recirculation of corrosion products.

Since the early 1960s, ferrous sulphate has been widely used in the power industry to maintain and supplement the natural protective film on copper alloy condenser tubes [8]. The mechanism of this has been investigated by a study of the corrosion characteristics of these materials under controlled erosion–corrosion conditions using the jet impingement technique. In this, flowing, once-through seawater impinges on the specimen surface from a 2 mm orifice 2 mm from the specimen surface at a jet speed of 5 ms^{-1} with 3 per cent air bubbles added. It was found that materials with an already fully formed, coherent iron-rich film, identified as lepidocrocite, γ-FeO·OH, exhibit a very high resistance to erosion–corrosion, but films produced by intermittent injection under active erosion–corrosion conditions are not fully protective; indeed, the film does not form at the impingement site and erosion–corrosion continues, unabated [2].

An electrochemical investigation into the mechanism for the protection of copper alloy condenser tube materials on film formation by ferrous sulphate has been completed [9]. It was found that there was a significant change in corrosion potential of nearly 50 mV in the anodic direction on changing from normal water speeds of $2\ ms^{-1}$ to erosion–corrosion water speeds of $8.5\ ms^{-1}$. The formation of an iron-rich film by continuous injection under these conditions caused the electrode potential to move in the cathodic direction. A determination of the anodic and cathodic polarization characteristics was carried out by potentiodynamic methods when it was found that formation of a fully coherent iron-rich film was accompanied by considerable polarization of both the anodic process (the loss of metal by ionization), and the cathodic process (the reduction of dissolved oxygen).

Two current theories for the mechanism of lepidocrocite film formation [11,12] suggest that the bonding is due to the electrophoretic migration of a charged colloid of lepidocrocite to the cathodic sites, which carry an opposite charge. These electrochemical studies indicate that as coverage of such sites is achieved then progressively smaller current densities are required to produce a given cathodic polarization, i.e. progressively smaller areas are available for electron transfer to occur between the diffusing oxygen and the ionizing metal. The suppression of these reactions leads to the suppression of that part of the corrosion reaction which occurs by this mechanism.

Polarization of the anodic process is fully consistent with anodic inhibition or passivation due to an insoluble oxide forming a barrier against the passage of ions to and from the reaction site. These results strongly support the views on the mechanism of erosion–corrosion expressed by Venzcel and his co-workers [13] who suggest that the rate controlling process under high velocity and erosion–corrosion flow conditions for copper and copper alloys is the ionization of the metal, i.e. the anodic process. This would imply that any treatment which retards the anodic process should also reduce the incidence of erosion–corrosion. This work fully substantiates such an implication and is in accordance with recent work on anodic polarization measurements under erosion–corrosion conditions [14].

It has been demonstrated that cathodic protection may be fully effective on the control of erosion–corrosion [2], but a study of the attentuation of polarization down a condenser tube shows that, given a fully polarized tube/tubeplate interface, concomitant with a uniformly and fully protected ferrous water box (polarized to -850 mV (Ag/AgCl)), polarization of the tube decayed to non-protective open circuit potential levels within a distance approximating to six tube diameters. Cathodic protection may therefore be effective in the control of inlet end erosion–corrosion but totally ineffective further down the tube.

NEW MATERIALS EVALUATION AND OPTIMIZATION

The evaluation of new materials for seawater-cooled condenser application has been in progress for many years. The possibility of using stainless steels for the construction of turbine condenser tube nests for CEGB plant was originally considered for the following reasons:

(a) to eliminate a potential source of contamination of the condensate and hence the feedwater by copper and other metals leached from conventional copper based tube alloys;
(b) to improve the integrity of condensers by permitting welding of tube/tubeplate joints;
(c) to replace the established tube alloys where technical or economic advantages could be shown.

The first research programme on stainless steels for condenser tube application was carried out by CEGB SE Region between 1965 and 1971 [15]. In this work, four materials were assessed, AISI types 316, 316L, 304 and 304L (Table 9.1).

Table 9.1 — Austenitic stainless steels

	Cr	Ni	Mo	C
Type 304	18–20	8–12	—	0.08 max
Type 304L	18–20	8–12	—	0.03 max
Type 316	16–18	10–14	2–3	0.08 max
Type 316L	16–18	10–14	2–3	0.03 max

After extensive laboratory and power station trials in clean seawater and polluted estuarine conditions, they were found to be unsuitable for condenser tube service for the following reasons:

1. They were susceptible to crevice corrosion at interfaces such as rubber-metal joints and at tube/tubeplate expansions.
2. They were susceptible to pitting corrosion, this being in general unrelated to particular defects. Pitting corrosion sometimes initiated at welds, but in general, the welded areas did not behave in a significantly different manner from the rest of the tube material which behaved in an unpredictable manner. All materials were fully resistant to erosion–corrosion as assessed in the standard jet impingement test [2].

Work on assessment of stainless steel materials continued on an ad hoc basis until 1978 when it became clear that developments in the steel industry had led to an availability of new alloys for which great improvements were being claimed [16,17] and a more formal programme of laboratory assessment of these materials was initiated.

To date, five materials have been assessed (Table 9.2). Of these, two were austenitic (AL-6X and Avesta 254 SMO) and three were ferritic (AL 29-4C, Sea Cure and Monit 2502).

The evaluation of these materials in tube or plate form using natural, once-through flowing seawater is more time consuming than the acidic ferric chloride test

Table 9.2 — Some recently developed stainless steels

	Cr	Ni	Mo	Mn	Ti	C	Alloy type
Allegheny Ludlum Al-6X	20.25	25.5	6.25	1.5		0.025	Austenitic
Avesta 254 SMO	20	18	6	0.5	0.4	0.02	Austenitic
Nyby Monit 2502	25	4	4	0.5		0.025	Ferritic
Trent Tube 'Sea-Cure'	26	1.7	3.4	0.5	0.5	0.02	Ferritic
Allegheny Ludlam 29–4C	29	—	4			0.02	Ferritic

(ASTM G-48) but the penalty of the increased time under test before meaningful results are obtained is considered acceptable in view of the service conditions anticipated.

Standard 25.4 mm od condenser tubes 1880 mm long have been used in a series of condenser tube rig trials. In each rig, twelve tubes were fixed in a plastics tube plate and nine rigs were available. Seawater flow was maintained at 1.8 ms^{-1} and certain rigs were operated with Taprogge on-load sponge ball cleaning or under syphonic conditions. Crevice corrosion may occur at the rubber O-ring seal in the tubeplate or at the plastics wedge-shaped turbulence raisers fitted in the tubes to simulate erosion–corrosion conditions.

The resistance to crevice corrosion was assessed using plate specimens and the INCO multi-crevice test jig [18]. The jig consists of two castellated plastics nuts, each consisting of twenty plateaux on the castellated surface. These are mounted on a screwed spindle. The nuts are machined so that there is no restriction to the flow of seawater between the castellations when the jig is assembled on the test plate. Three jigs may be accommodated, and each assembly, therefore, has 120 potential crevice corrosion sites, each of identical geometry.

Trials have been completed which extended to 2 years under these conditions and specimens were regularly inspected. Replicate crevice corrosion assemblies were removed, disassembled and examined at 6-month intervals. The resistance of the materials to erosion–corrosion [2] was also assessed.

In the condenser rig trials, the austenitic materials, AL-6X and Avesta 254 SMO, showed the most severe crevice corrosion on the outer surfaces of the tube, the corrosion on the AL-6X having started at the outlet end and continued down the tube as the result of the crevice conditions at the tube/tubeplate interface. The corrosion on the 254 SMO tube, on the other hand, had started at the outer edge of the tube/tubeplate joint and was confined to a narrow band at this point; this was probably due to leakage of seawater through the joint and the leakage to atmosphere clearly being oxygen saturated and promoting oxygen concentration cell conditions at the inter-face. Two of the ferritic steels, Sea-Cure and Monit, also had areas of very shallow crevice corrosion on the outside associated with the plastics tubeplate but the depth and extent of the pitting was very much smaller than that observed in the austenitic specimens. No external crevice corrosion had taken place on the 29–4C tube.

When the tubes were sectioned longitudinally it was found that there was no sign of pitting corrosion visible to the naked eye and it was necessary to make a detailed examination using a binocular microscope before any pitting was observed. It was found that all the materials had suffered pitting corrosion, but only very slightly in each case. There was no evidence of the pitting taking place preferentially at the welded seams, nor at the plastics turbulence raisers.

For plate specimens with the crevice corrosion test assemblies, it was found that of the materials examined, only the AL-6X showed pitting which could be definitely associated with the crevice corrosion test jig; some pitting was observed in the 254 SMO material and some very low levels of pitting in the three ferritic materials. The pitting was not observed until the specimens had been exposed for the full 2-year period. This crevice corrosion was also very shallow ($<5\mu$) and as only one specimen of each material suffered the effect, little distinction can be made between them in this respect. It was observed that the crevice corrosion was restricted to the extreme edges of the castellated nuts. This is the area where the maximum electrochemical defect would be anticipated as it delineates the cathodic, fully oxygenated areas and the anodic areas under the crevice being in an oxygen-restricted condition. The narrow band of crevice corrosion (no wider than $c.\ 75\mu$) after 2 years' exposure clearly demonstrates the very high resistance of these materials to crevice corrosion. On a mechanistic viewpoint, it demonstrates that crevice corrosion in less resistant materials probably originates in this form and propagates rapidly as the protective film breaks down. For a condenser tube under operational conditions, it is most unlikely that crevice corrosion conditions, resulting either from biological fouling or the lodgement of other foreign matter, would be expected to continue for this period of time, or even allowed to initiate, as control by conventional biofouling prevention methods and normal condenser inspection and maintenance routines would detect and rectify blockages by debris.

It was readily established that all these materials were fully resistant to erosion–corrosion, both after the standard jet impingement test and under the active erosion–corrosion conditions at the plastics turbulence raisers in the condenser tube rig trials which took place under syphonic conditions.

The very low levels of pitting due to either crevice corrosion or any other pitting mechanisms indicates that some considerable progress has been made in the specifications of stainless steels for seawater services [19]. It is clear, however, that certain claims that these materials may be 'totally immune' from crevice corrosion and pitting corrosion cannot be supported by the results of this work.

The pitting which was induced by this evaluation was very shallow and in no case had tube perforation taken place. The results to date do confirm results presented elsewhere [20] in that the high chromium, high molybdenum ferritic material show some superiority over the high molybdenum austenitics.

The only condenser tube material which has proved entirely satisfactory under all experimental test conditions, condenser simulation rig trials, and over recent years, actual service conditions, is titanium.

Titanium was reported to have a very high resistance to chloride environments

[21] and was finding wide application in desalination equipment [22]. The performance here has been such that life expectancies of up to 30 years are confidently expected. Laboratory studies have indicated a susceptibility to crevice corrosion in salt solutions at elevated temperatures [23] but there is no evidence to suggest that this mode of failure would occur in seawater at ambient temperatures.

Trials at the Brighton Seawater Corrosion Laboratory have subjected titanium condenser tubes to extended trials (6 months) in condenser tube rigs at normal water velocities (1.75 ms^{-1}), very high velocities (5.2 ms^{-1}) and syphonic conditions (2.4 ms^{-1}, 457 mm mercury vacuum). Of the twelve tubes in each rig, six were fitted with plastics turbulence raisers. These conditions are probably representative of the most severe and aggressive that any condenser tube would be expected to encounter under service conditions. No corrosion occurred, and the trials fully supported other observations that the resistance of titanium condenser tubes to corrosion by normal and high speed seawater is exceptionally good. Any susceptibility to crevice corrosion in condenser tube materials is normally shown up in these rigs at the tube/tubeplate seal; no such corrosion was observed.

Erosion–corrosion trials using titanium condenser tube samples have been carried out as follows:

1. Seamless titanium (UK manufacture); 60-day trial at 10 ms^{-1} jet speed+3 per cent air bubbles.
2. Seam-welded titanium (UK manufacture); 56-day trial at 10 ms^{-1} jet speed+3 per cent air bubbles.
3. Seam-welded titanium (USA manufacture); 56-day trial at 10 ms^{-1} jet speed+3 per cent air bubbles.

In the case of 2 and 3 the impinging jet was directed at the welded seams as it was considered that these were the areas most susceptible to possible failure.

In both the erosion–corrosion trials and the condenser tube rig trials described above, the flow conditions were such that severe corrosion, pitting and perforation would be readily induced in conventional copper-based condenser tube materials. That no such corrosion was observed to take place lends strong support to the view that titanium may be specified with a high degree of confidence for modern high-merit plant where condenser integrity is of such vital importance.

This laboratory work has been confirmed by condenser tube rig trials operated at proposed new power station sites. Three trials have been completed at Grain, Heysham and Ince 'B'; of the condenser tube materials included in the trials, only titanium was completely free from all evidence of corrosion and was selected as the condenser tube material at all three sites [24].

This work confirmed titanium as the only material which has shown a very high resistance to abrasion damage from suspended solids and a total immunity to erosion–corrosion, crevice corrosion and pitting corrosion in a wide variety of environmental and experimental conditions. Titanium has been installed, in trial batch quantities, at both seawater and estuarine-cooled CEGB condensers since 1959. Sytematic NDT surveillance since then has shown that no deterioration has taken place. The first complete condenser retube was carried out in 1972 and current CEGB policy is to specify titanium for the condensers and heat exchanges of all new plant where seawater or saline estuarine water is to be used as coolant [20]. Seam-welded 0.7 mm (0.028 in.) wall thickness, 25.4 mm (1 in.) od tube in titanium Grade 2 is the current standard for both new condenser construction and condenser retubing tasks. 0.5 mm wall thickness 25.4 mm od is also available. At the time of writing (February 1988), over 14 000 MWe of plant in the CEGB utilizes titanium as the condenser tube material.

The change to a thinner wall, lower density tube material has emphasized the need to reduce machine-induced vibrations which may cause damage by fretting, fatigue or impact. This is a complex phenomenon and Coit [25] has evolved the following formula:

$$L = \frac{9.5 \ (EI)^{0.25}}{\rho V^2 D}$$

where L=maximum allowable span length, E=modulus of elasticity of tube material, I=neutral axis moment of inertia, ρ=steam density, V=steam velocity, and D=tube outside diameter, and it is therefore to be seen that the maximum allowable distance between support plates is primarily a function of the elastic modulus of the material and the moment of inertia which relates to tube wall thickness. The modulus of elasticity of titanium is lower than that of stainless steel and copper alloy tube materials and this leads to the maximum distance between support plates being shorter for a condenser designed for titanium (Table 9.3).

Table 9.3 — Young's modulus of some condenser tube materials

	psi$\times 10^6$	GPa
Sea cure	31.2	215.1
Monit	31.2	215.1
Al-6X	27	186
Al 29-4C	30	206.8
254 SMO	29.03	200
Titanium	15.5	106.9
Cu/Ni 70/30	22	151

In his paper, Coit presents data which show that the maximum allowable span length for a titanium tubed condenser in 0.7 mm thick tubing is 87 per cent of that for brass tubed condensers. When titanium is selected as the condenser tube material at the design stage, tube plate support spacings are designed to minimize vibration effect. In retubed condensers, the replacement of tubeplate supports is not economically viable. A system of mid-span damping using rubber tubes to suppress vibration is usually used in the UK [26] (Fig. 9.5).

Fig. 9.5 — Mid-span tube damping system in a re-tubed condenser.

Titanium, unlike many copper alloy materials, is non-toxic to marine organisms [21] and prevention of condenser biofouling to maintain effective heat transfer conditions is of paramount importance. This problem was addressed as part of a biofouling control programme [6] and the results are shown in Table 9.4, from which

Table 9.4 — Overall mean heat transfer coefficients ($Wm^{-2} K^{-1}$)

Untreated	Taprogge	Cl_2, 0.5 ppm 20 min/hr	Cl_2, 0.2 ppm continuous	New tubes
$\bar{x}$=3139	$\bar{x}$=3597	$\bar{x}$=3628	$\bar{x}$=3640	$\bar{x}$=3936
σ=283	σ=93	σ=79	σ=63	σ=37
79.8%	91.4%	92.1%	92.5%	100%

it may be seen that the formation of biologically active slime films on the waterside surfaces of titanium condenser tubes can lead to a substantial reduction in overall heat transfer coefficient when compared with new tubes. The formation of these films is effectively inhibited by both chlorination regimes assessed in this work, 0.2 ppm continuously available chlorine and 0.5 ppm available chlorine for 10 minutes every hour. The action of recirculating sponge rubber balls in the Taprogge system [7] is similarly effective in the removal of the films formed by non-chlorinated seawater on titanium condenser tube surfaces.

There is a significant galvanic disparity between titanium and copper alloy tubeplate materials and corrosion of the latter will occur unless protective measures are adopted [27]. The galvanic corrosion process is accelerated by the highly turbulent conditions at the tube/tubeplate interface [28]. The complex aluminium bronzes have been shown to be most resistant to these effects but, where serious corrosion has developed or is anticipated, some corrosion-preventative measures to conserve the copper alloy tubeplates must be adopted.

Evidence is available to show that cathodic polarization in aqueous solution within a wide pH range can lead to titanium hydride formation [29,30].

In the context of an operational heat exchanger using seawater or saline estuarine water, the cathodic polarization required to achieve hydrogen evolution ($-700\,\text{mV}$, SCE) is available only from a cathodic protection system, and has been observed in a Japanese installation where impressed current cathodic protection was used to protect a brass tubeplate in a titanium tubed condenser from galvanic corrosion; the cathodic protection level was excessively high, -1.2 V (SCE) [31]. The effect is readily reproduced experimentally [32].

Some US experience of the protection of a titanium tubed condenser with an aluminium bronze tubeplate (ASTM B-171, CA 614) using impressed current cathodic protection has been reported [33] but the CEGB policy is to regard cathodic protection for this purpose as unacceptable. Polarization of the tubeplate at potentials more electronegative than -700 mV (Ag/AgCl) would present the tube/tubeplate interface with the hazard of hydrogen evolution. The possible formation of titanium hydride and consequent embrittlement of the titanium at these potentials, which are 150 mV electropositive to the potential required to protect the ferrous components, is regarded as an unacceptable hazard by the CEGB.

However, a recent report on US experience [34] describes the physical properties of tubes which had formed titanium hydride $\geqslant 50$ per cent wall thickness as a consequence of cathodic protection operating in excess of 1V (SCE). Tensile, flattening and hydrostatic pressure tests showed that the changes in mechanical properties attributable to hydride formation were insufficient to cause concern in view of the static loads anticipated. Measures were taken to control the levels of current applied in the cathodic protection system to maintain a potential of $-700\,\text{mV}$ (Ag/AgCl) to ensure non-hydriding conditions.

This level of potential, which was subsequently reduced to -650 V (Ag/AgCl), has been shown in some recent CEGB work to offer complete protection to the naval brass in a naval brass/titanium condenser rig operating under syphonic conditions ($2.4\,\text{ms}^{-1}$ water speed, 457 mm Hg vacuum) [35]. Work on the control of galvanic corrosion of a stainless steel tube/Muntz metal tubeplate assembly (a similar galvanically incompatible situation) indicates that the elimination of the galvanic current by cathodic polarization requires polarization to significantly lower polarization levels than this [36], confirming work describing the cathodic protection of copper alloy condenser tube materials [2].

The optimum solution to this problem is to eliminate the tube/tubeplate galvanic corrosion problem by using titanium tubeplates. The titanium tubes are welded to the tubeplate by the tungsten inert gas welding method with an appropriate joint preparation. CEGB work has confirmed the viability of this method to produce tube/tubeplate seam welded joints of high quality [37]. There are reports of its successful use in Japanese [38] and Swedish [39] power plant. The adoption of this system for future seawater cooled power plant clearly presents the possibility of a high integrity condenser where chloride ingress into the steam space by corrosion damage or leakage through mechanical joints is totally eliminated.

The most recent development in titanium condenser technology is in the field of advanced heat transfer surfaces. In these, the tubes are manufactured with a shallow spiral groove rolled on to both the inner and outer surfaces; these are colloquially referred to as 'roped' tubes (Fig. 9.6). The increased turbulence in the seawater has

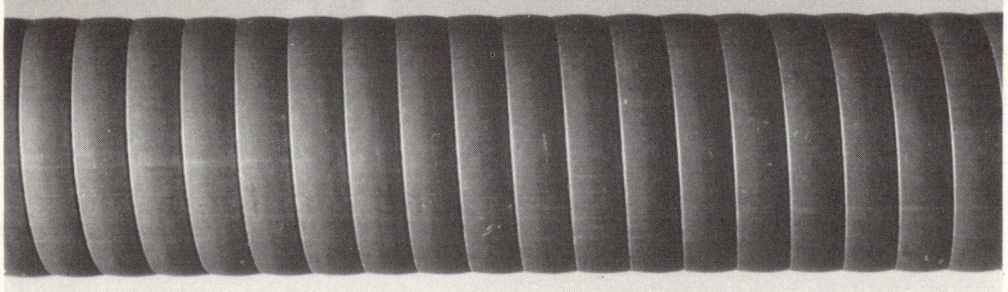

Fig. 9.6 — 'Roped' titanium tube.

no effect on the titanium in view of its total resistance to erosion–corrosion but there is a significant improvement in heat transfer efficiency. This is accompanied, on the steam side, by an improved drainage of condensate, both effects contributing to condenser efficiency. Although there is a concomitant increase in the cooling water friction factor, preliminary evaluation studies indicate that worthwhile modifications in steamside performance (possibly 2 to 3 mbar improvement in vacuum) may be achieved [40]. For the anticipated service conditions, waterside biofouling control is of paramount importance and studies at the Seawater Corrosion Laboratory have shown that conventional chlorination or Taprogge sponge ball cleaning will maintain these improved results [41].

ACKNOWLEDGEMENTS

This paper is published with permission of the Director of Plant Engineering (Nuclear), Operational Engineering Division, Central Electricity Generating Board.

REFERENCES

[1] Gilbert, P. T. Paper 194, *Corrosion 81*, Toronto, National Association of Corrosion Engineers, 6–10 April 1981.

[2] Heaton, W. E. *British Corrosion Journal,* **12** (1), 15, 1977.

[3] Page, G. G. *New Zealand Journal of Science,* **26**, 415, 1983.

[4] Page, G. G. *Anti-Corrosion Methods and Materials,* **14** (5), 13, 1967.

[5] Whitehouse, J. W., Khalanaski, M. and Saroglia, M. Symposium of Condenser Macrofouling Control Technologies: The State of the Art. EPRI CS-3343 (1983), Section 17.

[6] Heaton, W. E. Scottish Marine Biological Association 19th Scientific Meeting, Oban, 6–7 April 1982, 290.

[7] Eimar, K. and Renfftlen, R. Symposium on Condenser Macrofouling Control Technologies: The State of the Art. EPRI CS-3343 (1983), Section 13.

[8] Bostwick, T. W. *Corrosion,* **17** (8), 12, 1961.

[9] Heaton, W. E. *British Corrosion Journal,* **13** (2), 57, 1978.

[10] Ensign Plastics Ltd, British Patents 1247429, 1249596, 1344812.

[11] North, R. F. and Pryor, M. J. *Corrosion Science,* **8** (3), 149, 1968.

[12] Gasparini, R., Della Rocca, C. and Ionannilli, E. *Corrosion Science,* **10** (3), 157, 1970.

[13] Venzcel, J., Knutsson, L. and Wranglen, G. *Corrosion Science,* **4** (1), 1964.

[14] De Sanchez, S. R. and Schiffrin, D. J. *Corrosion Science,* **28** (2), 141, 1988.

[15] Baker, D. W. C., Heaton, W. E. and Patient, B. C. *Corrosion Science*, **12** (3), 247, 1972.

[16] Bond, A. P., Dundas, H. J., Ekerot, S. and Semchyshen, M. *Stainless Steels for Seawater Services, Paper 15*, Stainless Steel 77, Climax Molybdenum Company.

[17] Maurer, J. R. Stainless steel condenser tubes: economy, reliability, performance. Inco Power Conference 1977.

[18] Rowlands, J. C. *British Corrosion Journal,* **11** (4), 1958, 1976.

[19] 'Advanced stainless steels for seawater applications', *Proceedings of a Symposium held at the University of Piacenza*, 28 February 1980. The Climax Molybdenum Company, 1981.

[20] Streicher, M. A. *Materials Performance, 22* (5), 37, 1983.

[21] Cotton, J. B. and Downing, B. P. *Transactions of the Institute of Marine Engineers, 69*, 311, 1957.

[22] Fiege, N. G. and Kane, R. L. *Metals Engineering Quarterly, 7* (3), 28, 1967.

[23] Griess, J. C. *Corrosion,* **24**, 96, 1968.

[24] Heaton, W. E., Edgley, J., Andrews, E. F. C. and Patient, B. C. Paper C78/79, Institution of Mechanical Engineers Symposium *Cooling with Seawater*, London 15–16 May 1979.

[25] Coit, R. L. *Combustion,* **46** (2), 14, 1975.

[26] Hick Hargreaves & Co. Ltd, British Patent 1188564, 16 October 1978.

[27] Gehring, G. A. and Kyle, R. J. Paper 60, *Corrosion 82*, Houston: National Association of Corrosion Engineers.

[28] Case, B., Davies, J. O., Heaton, W. E. and Sketchley, J. M. *Colloquium on Choice of Materials for Condenser Tubes and Plates and Tubes and Tightness Testing*, Avignon, 22–24 September 1982.

[29] Phillips, I. I., Poole, P. and Shreir, L. L. *Corrosion Science,* **12**, 855, 1972.

[30] Phillips, I. I., Poole, P. and Shreir, L. L. *Corrosion Science,* **14**, 533, 1974.

[31] Sato, S. 'Corrosion and fouling of condenser tubes', paper presented to a

symposium *Water Chemistry and Corrosion in the Steam Water Loop of Nuclear Power Stations*, Scillac, 17–21 March 1980.

[32] Sato, H., Fuzukawa, T., Shimogor, K. and Tanabe, H. Hydrogen pick-up by titanium held cathodic in seawater, paper presented to the *Second International Congress on Hydrogen in Metals*, Paris, 6–11 June, 1977.

[33] Simon, P. D. Paper 77, *Corrosion 83*, Anaheim, 18–22 April 1983, National Association of Corrosion Engineers.

[34] Fulford, J. P., Schutz, R. W. and Lisenby, R. C. Paper 87-JPGC-Pwr-F, *Joint ASME/IEEE Power Generation Conference*, Florida 4–8 October 1987.

[35] Heaton, W. E. and Davies, J. O. Unpublished work, 1986–87.

[36] Gehring, G. A. and Kyle, R. J. Paper 60, *Corrosion 82*, Houston, National Association of Corrosion Engineers.

[37] Morgan-Warren, E. J. Welded titanium condensers for power plant, paper presented to *4th International Conference on Titanium*, Kyoto, Japan, May 1980.

[38] Saburai, K., Itabashi, Y. and Konatsu, A. Large titanium-tubed condensers with welded tube to tubesheet joints, paper presented to the *America Power Conference*, Chicago, 21–23 April 1980.

[39] Multer, I. and Hedström, M. Experience of titanium in turbine condensers, Paper 84-JPGC-Pwr-17, *Joint ASME/IEEE Power Generation Conference*, Toronto, 1–4 October 1984.

[40] Rowe, M. Power plant condensers — recent CEGB experience, Paper presented to *Institution of Chemical Engineers Symposium Series No. 75*, Manchester, 22–23 March 1983.

[41] Pearce, D. L. and Davies, J. O. Unpublished work, 1986–87.

10

Use of high alloy stainless steels in desalination systems

B. Todd

INTRODUCTION

Over the last 20 years, the use of desalination processes to produce potable and process waters from seawater and brackish groundwaters has grown rapidly, particularly in oil-rich desert areas such as the Middle East. Two processes predominate in this industry, namely Multi-Stage Flash (MSF) and Reverse Osmosis (RO). Most existing plants use MSF but in recent years the use of RO has grown (see Table 10.1) [1].

These processes are now major markets for the metal-producing industries.

Table 10.1 — World desalination plant capacity (millions of US gallons/day)

Process		1967	1972	1975	1977	1980
Distillation	total	216	325	447	760	1459
	MSF	153	226	352	638	1293
	% MSF	71	69	79	84	88
Membrane	total	5.1	23	77	218	468
	RO	0.1	4	45	167	390
	% RO	2	17	58	77	84
All processes		222	349	524	975	1922

Plants in operation or under construction on 1st January of the year (1980—30th June).

Although most of the alloy materials used are materials such as 90/10 cupronickel and Type 316L stainless steel, there are growing markets for high alloy stainless steels, particularly in seawater RO plants.

MULTI-STAGE FLASH

General

MSF is a distillation process where the incoming feed passes through heat exchanger tubes and is heated by vapour condensing on the outside (see Fig. 11.1) [2]. The condensate is the product from the plant and projects with an output of up to 200 million gallons per day have been built. The corrosive environments consist of:

1. Raw seawater — usually handled in 70/30/2/2 CuNiFeMn or titanium tubing.
2. Hot deaerated seawater — usually handled in 90/10 cupronickel or aluminium brass.
3. Incondensible gases such as CO_2, O_2, etc — Type 316L stainless steel is used.

In some cases, both raw seawater and incondensible gases are encountered in the same heat exchanger. These cases offer opportunities for high alloy stainless steels. Other possibilities are in control valves, pumps and acid-handling systems.

Vent system components

The gases from MSF plants are either condensed in shell and tube condensers or indirect contact condensers before venting to atmosphere. In the case of shell and tube condensers, the coolant is usually raw seawater which rules out the use of ordinary grades of stainless steels whilst the acidic condensates on the shell side rule out the use of copper-base alloys.

Titanium tubes have been used with some success in these units, but problems can arise with descaling. Mechanical descaling can lead to physical damage to the thin-wall titanium used to minimize cost. Acid cleaning can give difficulties with hydriding [3] of titanium. Some units have been tubed with a 6 per cent molybdenum stainless steel [4] — Avesta 254 SMO — and these have worked well. High alloy stainless steels are well suited to this market and are likely to replace titanium and copper-base alloys (still used and replaced periodically).

In the case of direct contact condensers, the vessels are often made of Type 316L stainless steels and, provided the vessel is well designed and fabricated, this works well. However, in some cases vessel internals hold pools of seawater and pitting and crevice corrosion can occur. Use has been made of high alloy materials — a 5 per cent molybdenum 25/20 NiCr stainless steel to line the lower parts of these vessels where seawater is likely to accumulate. In new projects, high alloy stainless steels are being specified for these vessels.

Another problem which arises with direct-contact condensers is carry over of seawater into the inter-vessel piping. This often results in stress-corrosion cracking of the Type 316L stainless steel piping used for these pipes. Piping in 25/20/4.5 NiCrMo stainless steels or Alloy 825 is now being used to counteract this problem.

Control valves

Control valves in MSF plants work under very arduous conditions as they sometimes handle hot brine, which with pressure reduction readily sets up cavitating conditions. These valves are often situated in concentric or eccentric reducers/expansion pieces so that velocities just upstream and downstream from the valves are high. The life of carbon steel or coated pipes is very short under these conditions and copper-base alloys can sometimes suffer impingement corrosion. Some use has been made of 904L stainless steel in this application and it has worked well.

Pumps

MSF plants use several large pumps for circulating seawater, brine and product water. The most commonly used materials are Ni-Resist casings with Type 316 stainless steel shafts and impellers. There has been some use of higher-alloy materials in pumps, notably duplex stainless steels for shafts and impellers.

All stainless steel pumps are sometimes used with cathodic protection to avoid pitting during shutdowns. Alloy 20 has shown poor results in this application as could be predicted from crevice corrosion tests in seawater which show it to be little better than Type 316 stainless steel.

Wear-rings are also sometimes made from duplex stainless steels — their higher hardness when compared with austenitic stainless should be of benefit in this application.

Acid handling systems

Some MSF plants use sulphuric acid as an anti-scalant. A pumping and piping system is needed to handle the acid, which is usually 97 per cent concentration. Although carbon steel is suitable for low-velocity parts of these systems such as storage tanks, it corrodes at an appreciable rate in areas of high velocity or turbulence, such as bends in pipe systems. The life of carbon steel piping systems is limited and it is more economic to use stainless steels such as Type 316L. Where there is a risk of external corrosion, then materials, such as 25/20/4.5/1.5 NiCrMoCu, or Alloy 825 should be used.

Pumps in these systems should also be made of corrosion-resistant materials and Alloy 20 (CN-7M) is preferred for this application.

REVERSE OSMOSIS SYSTEM

General

Fig. 10.1 shows a flow diagram of a typical RO system. Early membranes could effectively handle only brackish waters, but modern membranes can treat seawater.

One basic requirement is that the system upstream from the membranes should be corrosion-resistant, as corrosion products can block and interfere with the process. Although non-metallic materials such as glass-reinforced plastic can be considered for low-pressure parts, it is necessary to use metals for most of the high-pressure parts; in this respect, seawater RO plants normally operate at pressures in excess of 70 bar. Stainless steels, because of their very low general corrosion rate, are well suited to this application, but care is needed in selecting the correct grade so as to avoid pitting and crevice corrosion without overspecifying and incurring unnecessary cost.

Table 10.2 gives a range of composition for 25 brackish waters [6] either in use or to be used for RO processes. A mathematical modelling technique [5] has been used [6] to define the most suitable grade of stainless steel for a particular chloride level in the feed.

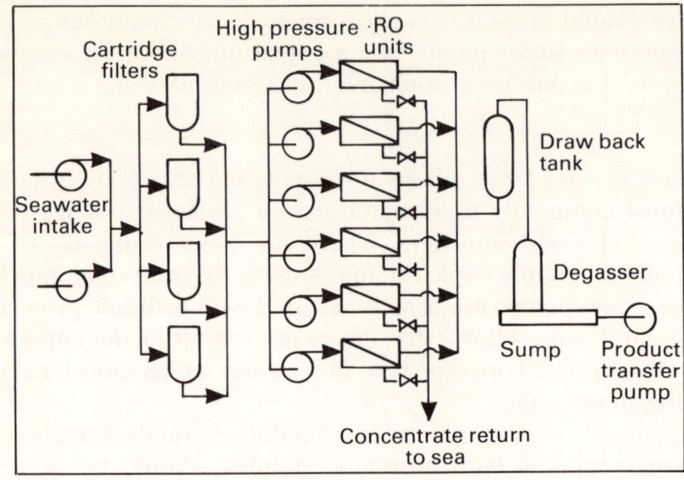

Fig. 10.1 — Flow diagram of a typical RO system.

Table 10.2 — Composition of 25 brackish waters

	Range (mg/litre, except pH)	
	Minimum	Maximum
pH	6.0	8.0
TDS	1930	20,700
Chloride	110	2,680
Sulphate	607	10,100
Sodium	400	5,250
Calcium	55	590
Magnesium	54	583
Bicarbonate	132	391
Oxygen	0	[a]
H$_2$S	0	[b]

[a]Fully aerated (level not normally given).
[b]Present (level not normally given).

Fig. 10.2 shows how this technique ranks various alloys in relation to chloride concentration. In this figure 'Exceptional' resistance means that no corrosion should occur, 'Very Good' resistance means that only under extreme crevice conditions

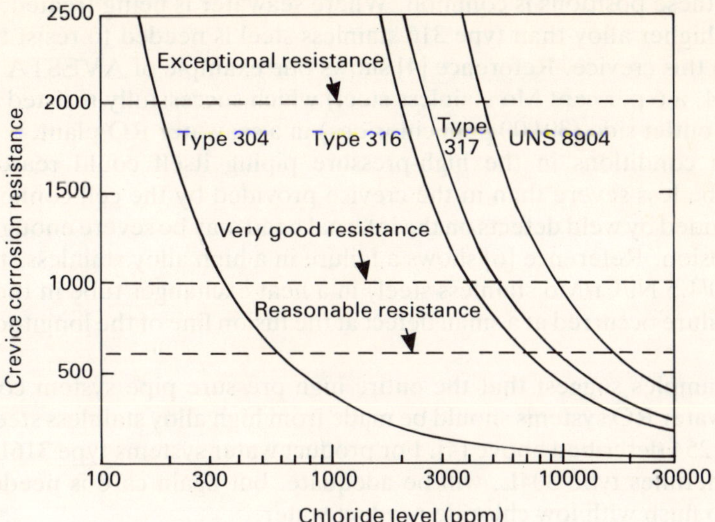

Fig. 10.2 — Predicted crevice corrosion resistance of a range of stainless steels in waters of varying chloride content.

would corrosion be expected, whereas 'Reasonable' resistance means that corrosion could occur in crevices normally encountered in the service. Thus, for seawater applications, alloys with higher resistance than the UNS8904 type are needed.

Pumps

During operation, high velocities maintain pasivity and pumps are normally made from standard grades such as Type 316 and its cast equivalents. However, because of the high pressures involved, the use of duplex materials of the 25 Cr 5 Ni 2.5 Mo type would be an advantage.

High pressure piping

In a large project, the high-pressure piping carries water from the pumps to banks of RO modules, where it is distributed through headers with connections to individual cells. There is thus a range of diameters used, the maximum being about 150 mm and the small individual cell connections 15–20 mm.

The piping is mostly fabricated by welding and, because of the sizes involved, this can normally only be performed from the outside. A careful welding technique is therefore needed to avoid serious defects on the inner bead of the weld. In our experience most corrosion failures in RO pipes are either in or close to welds, indicating that weld or heat affected zone (HAZ) defects have provided the initiation site for the corrosion.

The individual connections to the RO cells have to be demountable. Various devices are used to give a demountable joint, such as O-ring compression fittings, screwed connections, etc., but in all cases these provide a severe crevice and crevice corrosion at these positions is common. Where seawater is being treated, it is clear that a much higher alloy than type 316 stainless steel is needed to resist the severe conditions in this crevice. Reference [4] shows one example of AVESTA 254 SMO (UNS S31254, a 6 per cent Mo stainless steel) which successfully resisted corrosion on the brine outlet side (30 000 ppm chlorides) in a seawater RO plant.

Although conditions in the high-pressure piping itself could reasonably be expected to be less severe than in the crevice provided by the cell connectors, the crevices provided by weld defects on the internal bead may be severe enough to cause crevice corrosion. Reference [6] shows a failure in a high alloy stainless steel (UNS 8904, a 25/20/4.5 Ni/Cr/Mo stainless steel) in a heat exchanger tube in the Arabian Gulf. This failure occurred at a small defect at the fusion line of the longitudinal weld in the tube.

These examples suggest that the entire high-pressure pipe system connectors, etc., for seawater RO systems should be made from high alloy stainless steels such as the UNS S31254 described above [4]. For product water systems type 316L stainless steel, or sometimes type 304L, will be adequate, but again care is needed during shutdowns to flush with low chloride product water.

CONCLUSIONS

Both MSF and RO systems present good opportunities for the use of high alloy stainless steels. The RO process is still in an early stage of development, but in time is likely to become a major metal user for both standard and high-alloy grades of stainless steel.

REFERENCES

[1] Todd, B., 'Copper alloys in the desalination industry', Copper Development Association, *Copper '83* Conference, London, Nov. 1983.

[2] Oldfield, J. W. and Todd, B., 'Corrosion Problems Caused by Chlorination in the Desalination Industry', Chapter 11, this volume.

[3] Moroishi, T. and Miyuki, H., 'Hydrogen absorption of titanium in seawater at high temperature, *Proceedings 4th Intl Conf. on Titanium,* Japan, May 1980.

[4] Olsson, J. and Wallen, B., 'Experience with high molybdenum stainless steels in saline environments', *Proceedings 1st World Congress on Desalination and Water Re-Use,* Florence, 1983, Vol. 1, p. 241.

[5] Oldfield, J. W. and Sutton, W. H., 'New technique for predicting the performance of stainless steels in seawater and other chloride-containing environments', *British Corrosion Journal,* **15**, 31–36, 1980.

[6] Oldfield, J. W. and Todd, B., 'The use of stainless steels and related alloys in RO desalination plants', Institute of Marine Engrs/Bahrain Society of Engineers, *International Maritime Engineering conference,* Bahrain, Jan. 1984.

11

Corrosion problems caused by chlorination in the desalination industry

J. W. Oldfield and B. Todd

INTRODUCTION

Most desalination plants use the multi-stage flash (MSF) distillation process — about 70 per cent of installed capacity uses this process. This is essentially a heat exchange process in which seawater is caused to flash (boil) in a number of stages at progressively lower temperature and pressure (Fig. 11.1). The pressures in the plant are controlled by an ejector system, usually made from Type 316 stainless steel and in some plants serious corrosion problems have been experienced in this vent system. These problems have been caused by bromine being evolved from the seawater. Although most of the problems have occurred in plants where acid dosing is used to control scaling there have been some cases [2] in additive dosed plants. The characteristics of the corrosion are different in these two cases. In acid-dosed plants the stainless steel shows a multiplicity of fine pits which eventually cause perforation — cracking has not been found. In the additive plants corrosion is characterized by cracking associated with heavy general attack. Metallographically the cracking is characteristic of trans-granular stress corrosion in stainless steels.

The purpose of this chapter is to define the conditions under which these two types of corrosion occur and the mechanisms by which bromine can be evolved from seawater.

VENT SYSTEM IN CORROSION IN ACID DOSED PLANTS

The seawater entering an MSF plant is normally chlorinated to control marine growth. Most plants use continuous chlorination, leaving a low residual (0.1–0.2 ppm) at the plant outlet but some operators favour shock dosing and various regimes using several parts per million of chlorine are used.

The chemistry of the reactions of chlorine with the bromide ions (*c.* 70 ppm)

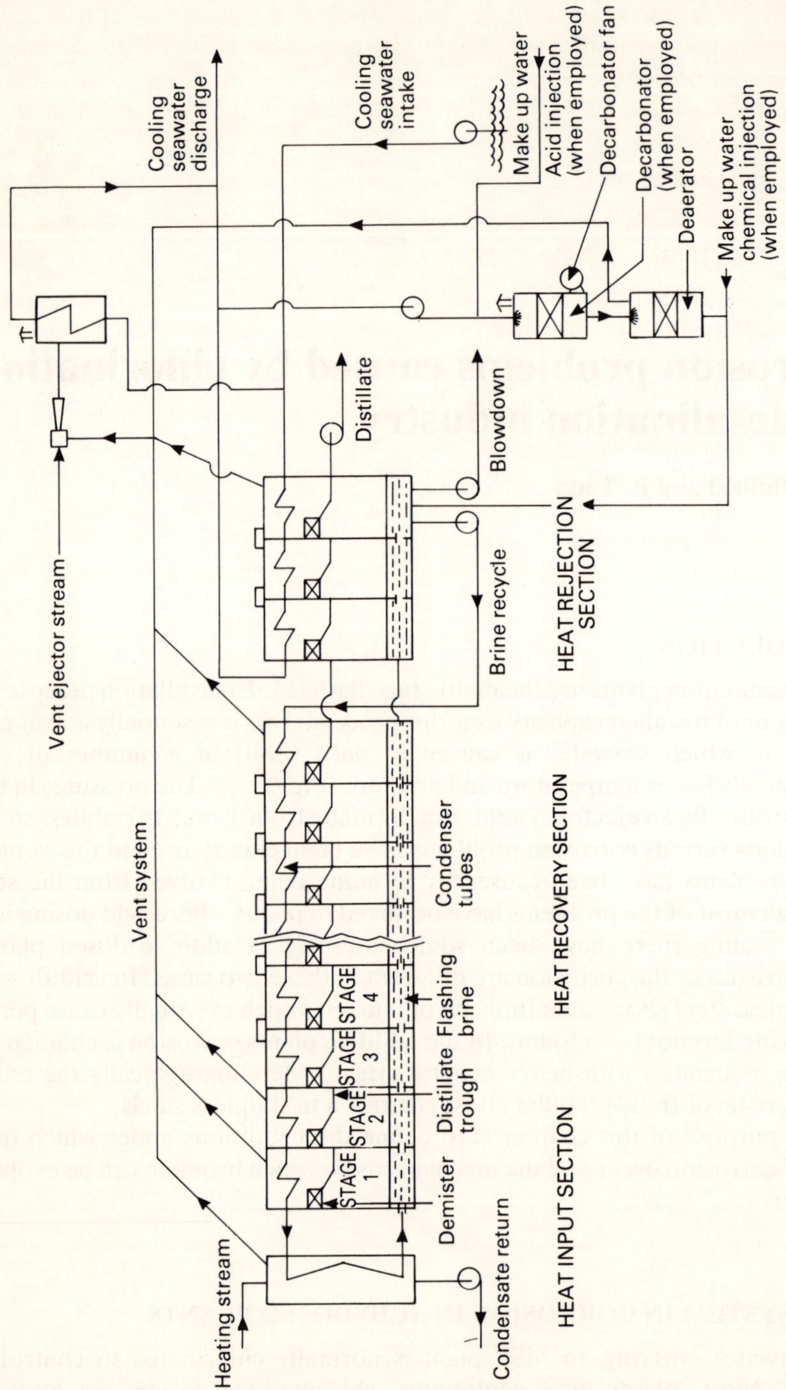

Fig. 11.1 — Simplified multi-stage flash (MSF) flow diagram.

normally present in seawater have been studied in Reference [1] which shows that the reaction goes virtually to completion in reducing residual chlorine and oxidising the bromides present to bromine.

Normal seawater has a pH of about 8.0 and at this level most of the residual bromine is present as hypobromous acid and hypobromite ion. The reactions involved are as follows:

$$Br_2 \text{ (gas)} \rightarrow Br_2 \text{ (aqueous)}$$
$$Br_2 \text{ (aqueous)} + H_2O \rightarrow HBrO + Br^- + H^+$$
$$HBrO \rightarrow H^+ + BrO^-$$

However, in an acid-dosed plant the pH of the seawater is reduced to about 4 in order to decompose carbonates and bicarbonates. The seawater is then passed through a decarbonator to remove carbon dioxide and then a deaerator to remove oxygen and other incondensible gases which could cause corrosion and reduce heat transfer in the plant.

Fig. 11.2 shows the effect of these changes in pH on the bromine compounds

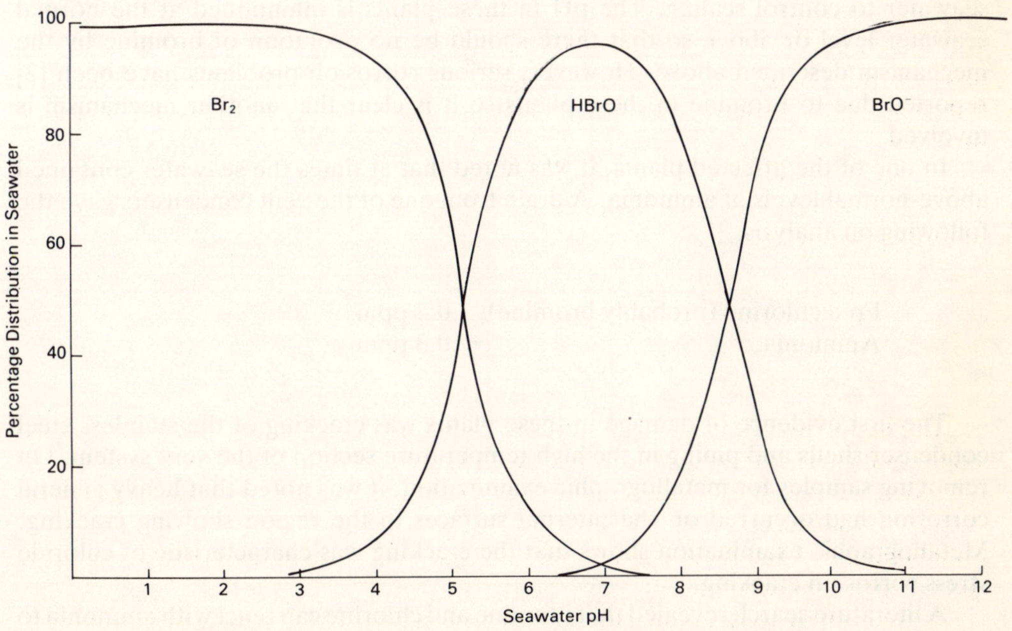

Fig. 11.2 — Distribution of bromine in seawater as a function of pH.

present from which it can be seen that, at pH 4, most of the halogen residual is present as bromine gas in solution. Although the pH rises to between 5 and 6 when the carbon dioxide is stripped out, even at these levels a significant proportion of the residual is present as bromine in solution.

When this seawater passes to the deaerator this bromine is stripped out with the other gases and passes into the ejector system. Also, as the mass flow of seawater through the deaerator is 1000–10 000 times the mass of the gases, the bromine is concentrated by this factor so that low levels in the seawater can become high levels in the vent system. Any condendsates from these gases will also have a much higher concentration of bromine than the seawater.

Fig. 11.3 shows the effect of seawater pH on the bromine concentration in the vent gases at three levels of residual chlorine (as measured in the seawater). Fig. 11.4 shows the bromine levels in condensates from the vent gases.

Work [3] on the effect of bromine on corrosion of stainless steels shows that attack is likely above 10 ppm. Fig. 11.4 indicates how by controlling residual chlorine and the pH at the deaerator inlet, the bromine level in the vent gas condensates can be controlled.

If for process reasons it is not possible to obtain the levels indicated in Fig. 11.4, then the feed should be dechlorinated [4] with sodium sulphite before entering the deaerator.

VENT SYSTEM CORROSION IN ADDITIVE PLANTS

In these plants polyphosphates or proprietary organic compounds are added to the seawater to control scaling. The pH in these plants is maintained at the normal seawater level or above so that there should be no evolution of bromine by the mechanism described above. However, serious corrosion problems have been [2] reported due to bromine in these plants so it is clear that another mechanism is involved.

In one of the affected plants, it was noted that at times the seawater contained above-normal levels of ammonia. A drain from one of the vent condensers gave the following on analysis:

> Free chlorine (probably bromine),　0.2 ppm
> Ammonia,　　　　　　　　　　　　　0.3 ppm

The first evidence of damage in these plants was cracking of the stainless steel condenser shells and piping in the high temperature section of the vent system. On removing samples for metallographic examination, it was noted that heavy general corrosion had occurred on the internal surfaces in the region showing cracking. Metallographic examination shows that the cracking was characteristic of chloride stress corrosion cracking.

A literature search revealed that bromine and chlorine can react with ammonia to form bromamines and chloramines as follows:

$$NH_3 + HOBr \rightarrow NH_2Br + H_2O$$

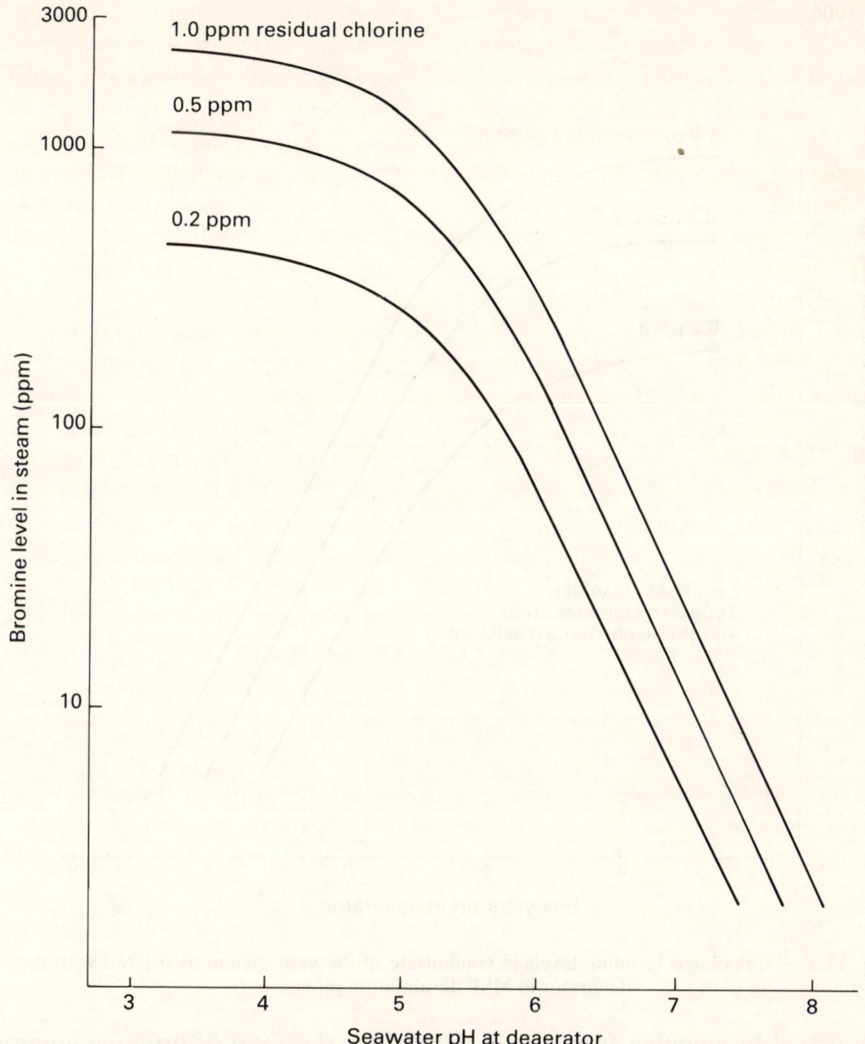

Fig. 11.3 — Calculated bromine levels in steam leaving deaerator in a MSF desalination plant for different residual chlorine levels.

$$NH_2Br + HOBr \rightarrow NHBr_2 + H_2O$$
$$NH_2Br_2 + HOBr \rightarrow NBr_3 + H_2O$$

Similar reactions occur with chlorine.

Experimental data indicate that the reaction time to form chloramines is of the order of minutes and would not be complete before the chlorine was reduced by the bromide ions present. Hence, in chlorinated seawater, bromamines would predominate.

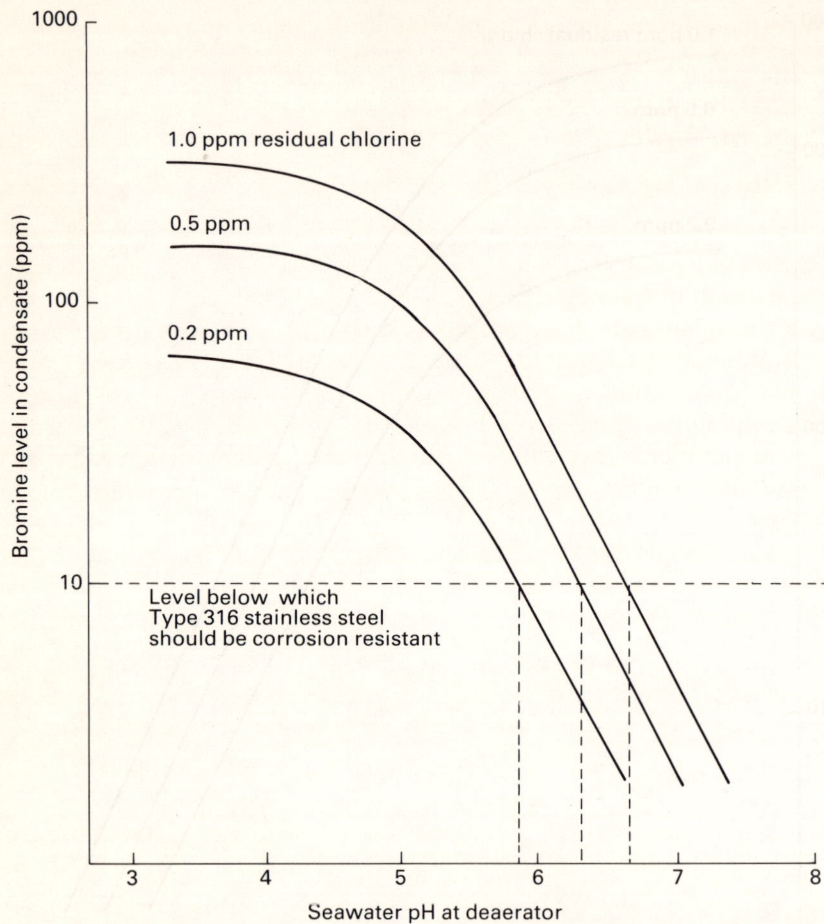

Fig. 11.4 — Calculated bromine levels in condensate in the vent system associated with the deaerator in MSF desalination plants.

The actual bromamine formed will depend on the ratio of bromine:ammonia present in the seawater. However, significant amounts of chloramine may form — Reference [5] indicates that at a 1:1 ammonia/chlorine ratio in terms of ppm (5:1 in molarity) about 10 per cent of the total oxidant present is monochloramine at pH 8 and 25°C.

Analysis of corrosion products also indicate significant chlorine evolution as shown by the following analysis:

Bromine,   3.3 per cent
Chlorine,   0.14 per cent

In the acid-dosed plants described earlier only a trace of chlorine is normally detected and the bromine to chlorine ratio is several thousand to one.

Once formed, bromamines (and chloramines) can decompose as follows:

$$4NH_2Br + H_2O \rightarrow N_2 + 3HBr + 2NH_3$$
$$2NHBr_2 + H_2O \rightarrow N_2 + 3HBr + HOBr$$
$$2NBr_3 + H_2O \rightarrow N_2 + 3HBr + 3HOBr$$

Formation of hydrobromic and hypobromous acid in the venting system is likely to cause corrosion of stainless steel.

In order to demonstrate the possibility of the above mechanisms for stripping bromine from seawater, laboratory experiments were set up in which a 3.5 per cent sodium chloride solution containing bromine and ammonia was degassed using argon as stripping gas. The evolved gases were absorbed in a pH 10.5 solution of caustic soda at ambient temperature. The experiments were carried out at 40°C and 90°C to simulate conditions at the high temperature and low temperature sections of an MSF plant.

The gases absorbed in the caustic soda were analysed using ultra-violet absorption.

Fig. 11.5 shows the results of these experiments which demonstrate that broma-

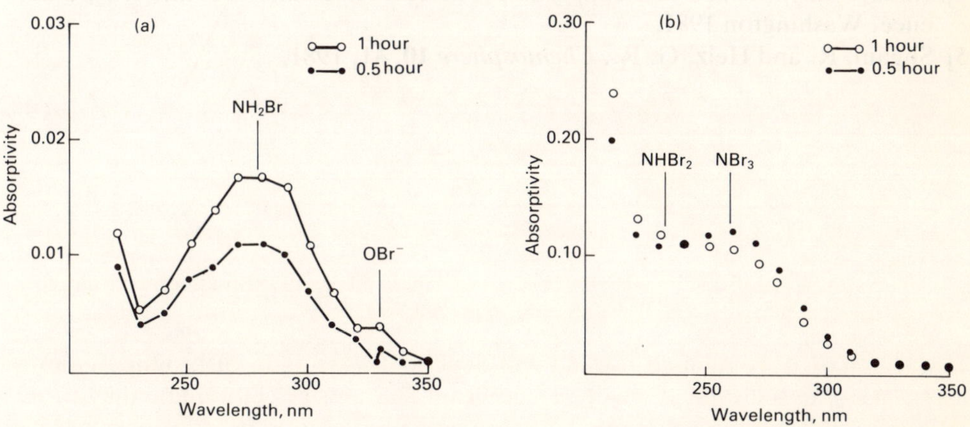

Fig. 11.5 — UV absorption spectra for bromamines stripped from a 0.56 M NaCl solution at (a) 40°C and (b) 90°C, and absorbed in a pH 10.5 NaOH solution at ambient temperature.

mines can be evolved from simulated seawater environments at both 40°C and 90°C. Also, as the times involved, up to 1 hour, are long compared with the time the seawater takes to flow through the plant, it is possible that bromamines would be evolved at any (or all) stages of the process. The fact that corrosion occurs only at the high temperature section of the plant may be because the bromamine decomposition occurs more readily there or because the concentration of bromine compounds increases as it passes through the vent system.

SUMMARY AND CONCLUSIONS

When seawater is chlorinated bromine compounds will normally be formed. In acid-dosed MSF plants when the pH is reduced any residual chlorine will be present partly as dissolved bromine gas. This gas can then be stripped out at the deaerator, causing corrosion in the venting system, unless care is taken to control the plant conditions.

For additive plants the presence of ammonia in the seawater can lead to the formation of bromamines and chloramines which can be stripped from the seawater to concentrate and decompose in the venting system. The cracking found in this case may be due to the chlorine carried over by the chloramines or it may be because one of the bromamine compounds is a stress corrosion cracking agent.

These failures emphasize the need to maintain chlorine residuals at as low a level as possible, consistent with controlling marine growth.

REFERENCES

[1] Oldfield, J. W. and Todd, B. *Desalination*, **38**, 233, 1981.
[2] Lee, W. S. W., Oldfield, J. W. and Todd, B. Corrosion problems caused by bromine formation in additive dosed MSF desalination plants, *1st World Congress on Desalination and Water Re-use*, Florence, May 1983, pp. 209–221.
[3] Morris, P. E. and Wendler, W. H., *Inco Internal Report Project No. 3093*.
[4] Nada, N., *National Water Supply Improvement Association in America* Conference, Washington 1981.
[5] Sugam, R. and Helz, G. R., *Chemosphere* **10**, 41, 1981.

12

Materials selection for electrochlorination systems

M. W. Upjohn
Electrocatalytic Ltd, Norman Way, Severnbridge Industrial Estate, Portskewett, Newport, Gwent, NP6 4YN

FOREWORD

The intention of this paper is twofold, firstly to increase the awareness of users of the material considerations and choices to be made when developing electrochlorinator systems and, secondly, to support the understanding of the users' requirements. The author has naturally drawn on his own company's experience obtained over many years. The paper describes the generator cell function and the equipment required for the handling of its products.

WHY CHLORINATE? — THE CHEMISTRY OF ELECTROGENERATION

General requirement for chlorine or hypochlorite solutions

The need for a supply of chlorine or hypochlorite in aqueous solution derives fundamentally from the ability of living organisms to colonize permanent installations through which continuous supplies of water are being passed. Waters from the sea or estuary are as prone to this colonization as fresh water systems and in some ways the fouling can be more serious for the salt water supplies. Growths of algae and particularly of molluscs become serious, resulting in restrictions to cooling water flow, and the overheating and fouling of the machinery involved.

It has been found over many years that the addition of chlorine to the water of the circulation system can inhibit the growth of marine organisms and consideration of solution concentrations and reactivities led to the realization that the principal germicidal action must be through the formation in solution of hypochlorous acid. It was also found that the characteristics of growth of the contaminating organisms determined that continuous addition of chlorine or hypochlorite over extended periods would be more effective than a massive single addition of the germicide, provided that an adequate minimum germicidal concentration was achieved. In particular, the control of molluscs demanded the use of a continuous dosing method.

Hypochlorite is now widely used in 'Waterflood' applications on various offshore oil production platforms. In this particular application it is necessary to pump vast quantities of water into the oil-bearing rock structure in order to force the oil out of the various production wells. As there is plenty of seawater easily available this is the obvious fluid to use, but unless the marine organisms in the seawater are very carefully controlled then these organisms will deposit in the pores of the rock strata and effectively reduce the oil flow.

The chemistry of electrogeneration of sodium hypochlorite

The electrochemistry of chlorine has been reviewed by Mussini and Fiati [1] and Kelsall [2]. In general, the chemistry of the chlorine oxy-acids is very complicated, as can be seen from the thermodynamic description of the chlorine-water system, described by Pourbaix [3] in terms of potential — pH diagrams. However, the basic reactions involved in the electrogeneration of sodium hypochlorite were formulated by Foerster and Muller [4] at the turn of the century.

$$\text{Anode: } 2Cl^- \rightarrow Cl_2 + 2e^- \tag{i}$$

$$\text{Cathode: } 2H_2O + 2e^- \rightarrow 2OH^- + H_2 \tag{ii}$$

$$\text{Solution: } Cl_2 + H_2O \rightarrow HOCl + Cl^- + H^+ \tag{iii}$$

$$HOCl \rightleftharpoons H^+ + OCl^- \tag{iv}$$

The chemistry and physics of the processes are quite complex, the electrochemisty being coupled to a chemical reaction in the anodic reaction layer, whereby the electrogenerated chlorine is immediately hydrolyzed to form hypochlorite. Both reactions (i) and (ii) have been studied by Landolt and Ibl [5] and some of the net conclusions of their work are given below.

The primary electrode reactions are the kinetically controlled discharge of chlorine and hydrogen, i.e. reactions (i) and (ii) respectively.

The chlorine evolved is hydrolysed by a chemical reaction in the reaction layer close to the anode surface. A study of this reaction by Spalding [6] has shown the chlorine to react with both water molecules and hydroxyl ions. From pH 3 to 10.5 reaction (iii) predominates and thus it may be assumed that at the pH values encountered in sodium hypochlorite electrogenerators, the concentration of dissolved elemental chlorine is negligible.

Landholt and Ibl [5,7] have proposed that part of the OCl^- formed was used to buffer the anodic reaction layer, as the $HClO/OCl^-$ equilibria is both pH and Cl^- concentration dependent, such that chlorine hydrolysis could occur at a sufficient rate. At high chloride concentrations (35.5 g/l) the hydrolysis reaction is interpreted as taking place at a mean distance from the anode greater than the diffusion layer thickness. Further studies have been conducted by Jaksic *et al.* [8].

Several subsidiary reactions [9–11] are theoretically possible, however, under the general conditions of operation of a hypochlorite cell. These are of negligible importance with the exception of the kinetically controlled decomposition of water with oxygen evolution at low hypochlorite concentrations ($HOCl$ and OCl^-). This reaction leads to the major loss of current efficiency but its effect may be minimized by judicious selection of an anode material with a high oxygen overpotential and low chlorine overpotential. Recent comprehensive studies on this subject of loss reaction have been performed by Lartey [12], Heal [13], Booth [14] and Kuhn [15].

Cell design criteria

The mixing by turbulence of the primary anodic (chlorine), and cathodic (hydroxide) reaction products within the cells will increase the rate of hydrolysis, reaction (iii), by which hypochlorite is formed. Should the cell be of bipolar design, as is the 'CHLOROPAC' cell, then one has the further advantage of having the potential for scale-up in terms of both current and voltage, with the advantage of reduced power supply costs by scaling up in terms of voltage rather than the current alone.

The choice of design is much wider with a bipolar cell as external connections are made to two feeder electrodes across which a sufficient voltage is applied such that the potential gradient in the solution produces anodic and cathodic areas on each intermediate electrode, i.e. electrical connection between the bipolar elements is made through the electrolyte. This allows small interelectrode gaps to be used. The design of a tubular cell overcomes the problems facing the bipolar 'Stack Type' parallel plate electrodes (Fig. 12.1) since this design tends to overcome the current by-passing the electrodes. A recent review on the problems facing parallel plate bipolar stacks is by Booth and Kuhn [16].

A problem facing plate-type hypochlorite electrogenerators has been blockage, during seawater electrolysis, by $Mg(OH)_2$ deposits, nucleated at cathode surfaces where hydrogen evolution leads to a high local pH. To overcome this, high electrolyte flow rates have been used at slightly increased pumping costs [17]. It may be seen that under normal operating conditions 'CHLOROPAC' tubular type cells operate at a Reynolds number in excess of 25 000 and this overcomes the problem. Thus, the practical problem of precipitation of hydroxides at cathode surfaces is evidently amenable to solution by appropriate cell design.

Tubular design philosophy

These cells were designed as flow-through undivided cells in which the saline water is subjected to electrolysis as the electrolyte flows through a series of these cells. The salt water flows through an annular opening at a velocity of some $3\,\text{ms}^{-1}$. This velocity will continuously flush out precipitates of insoluble salts normally found in seawater or other brackish waters.

Several authors have described the tubular cell: Houghton and Kuhn [18], Connolly [19], Lartey [12], Kuhn [20, 21], Anderson and Lamb [22], Kelsall [2] and White [23].

A cell (Fig. 12.2) is seen to be constructed of three titanium tubular shells. The outer tubular shell is electrically split halfway along its length, one half is the anode, the other being the cathode, both insulated by plastics spacers, while the inner tube requires no electrical connections — a floating bi-pole. The outer shells are placed axially in line and connected by their flanges with the insulating plastics spacers to form a smooth bore pipe internally. The inner shell is seen to be smaller in diameter and longer than an outer shell and thus may be placed inside the outer shells to complete the cell.

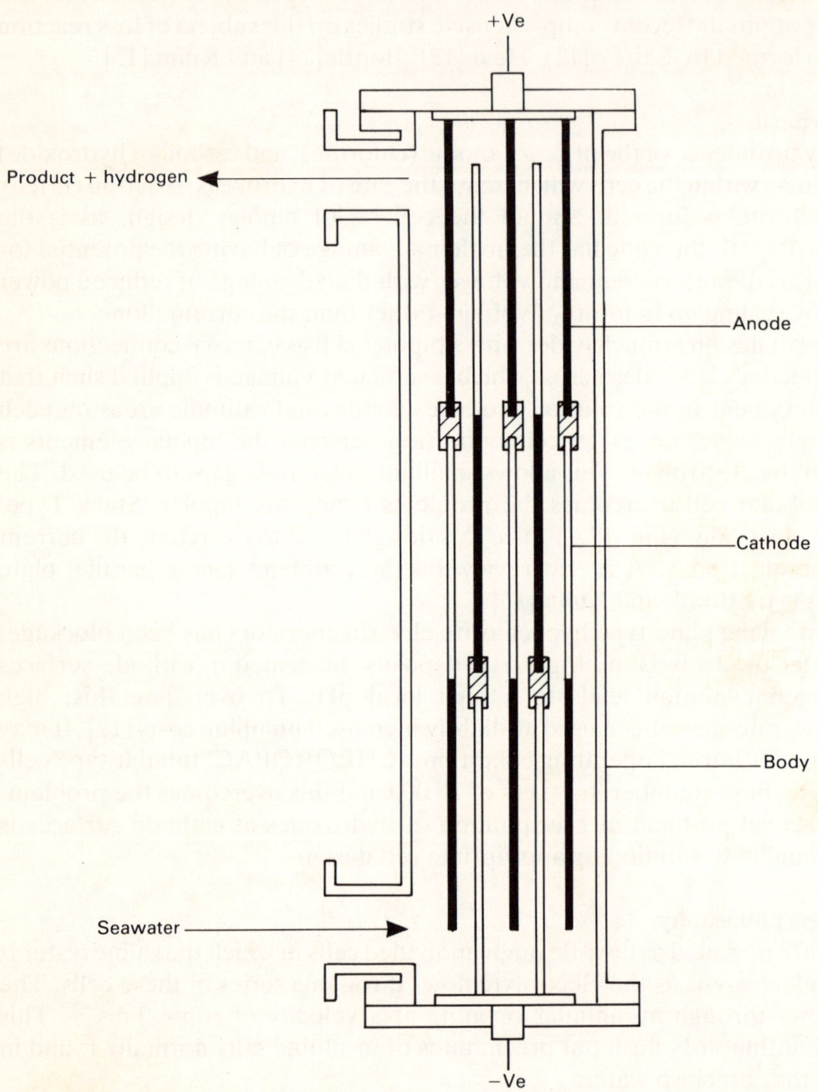

Fig. 12.1 — Plate cell design.

The internal surface of the outer anode is titanium, coated with platinum, and it is from this surface that the generated chlorine is immediately hydrolysed to sodium hypochlorite close to the diffusion layer. The internal surface of the outer cathode is bare titanium. The outside of the inner tubular shell adjacent to the outer anode receives electric current as a cathode (titanium surface only) and that half adjacent to the outer cathode is platinized titanium and acts as an anode. Here the current passes

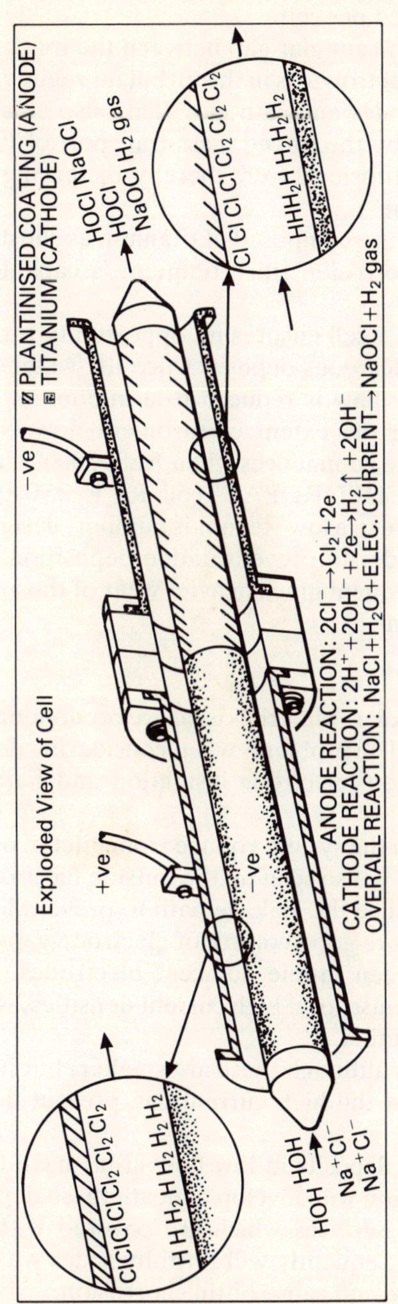

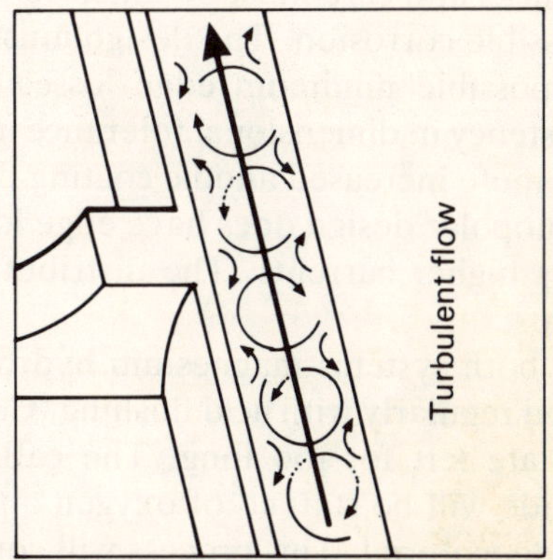

Fig. 12.2 — Tubular cell design.

from the anodic region of the inner tubular shell to the inner surface of the outer cathode. Thus, two current passes are achieved per cell.

The salt water electrolyte flows through the annular gap between the inner and two outer tubular shells. The flow stream of electrolyte is in the turbulent regime and thus mixes the products produced at the anodes and cathodes. This also aids the production of sodium hypochlorite solution by the forced mass-transport of OH^- ions from the cathode to near the anode diffusion layer, thereby increasing the hydrolysis of the anodically generated chlorine.

Since the inner and outer shells of the cell are composed of titanium as substrate they are extremely resistant to salt water corrosion since titanium is capable of forming a protective oxide coating.

The tubular design of the 'CHLOROPAC' cell ensures no opportunity for cell blockages by the $Mg(OH)_2$ and $Ca(OH)_2$ calcareous deposits since the water flow over the electrodes is turbulent and laminar flow is reduced to a minimum. The parameter used in cell design to determine the extent of turbulent flow is the Reynolds Number (Re). The turbulent regime commences when Re is 3000. Under normal operating flows and conditions a 'CHLOROPAC' cell Re is 25 000. In systems where the cell design parameters relate to a low Reynolds number, designers have to resort to high water velocities in order to prevent undue deposition, i.e. 'forced hydrodynamic' conditions, with consequent much heavier wear of the anode surfaces and subsequent loss of precious metal.

Plate type cells

There are two main types, monopolar and bipolar. Materials used are predominantly titanium, Hastelloy, and in some cases steel. The problems when considering design include anodic dissolution, cathodic corrosion, electrode insulation and Faraday efficiency and are similar for the two types.

When considering the bipolar stack design the system voltage from inlet to outlet may be up to 40–50 volts D.C. This potential is also seen on the outside faces of the plate stacks and care must be taken to minimize the leakage path to prevent losses and possible corrosion. The design must ensure good control of electrode gaps and where possible minimum edge losses between in-line adjacent electrodes. Any inconsistency in dimensional tolerance may cause local high current densities, which will promote increased anodic coating dissolution.

Monopolar design does have edge losses, although these are smaller, but it also requires higher currents. The distribution of the high current can present design problems.

For both systems magnesium hydroxide deposits in low flow areas have to be removed regularly with acid flushing. Corrosion will develop beneath these deposits if they are left for too long. The cathodic surfaces which are covered with the hydroxide will be starved of oxygen and consequently will form hydrides with the electrode material. This process will continue and cause pitting corrosion.

PROCESS REQUIREMENT

The basic equipment, instrumentation and control requirements for an electrochlorination system are listed below:

(a)   Seawater supply pumps and pressure indicator.
(b)   Seawater flow indication and low flow protection.
(c)   Filtration of the seawater.
(d)   Piping and valves suitable for the seawater.
(e)   Hypochlorite generating cells.
(f)   Product high temperature switch.
(g)   Generator low water level switch.
(h)   D.C. power source for cells.
(i)   Control panel with all required safety interlock and process control.
(j)   Hydrogen removal/storage tanks.
(k)   Level switches and controllers for tank.
(l)   Dilution air fans and air flow switches.
(m)   Dosing pumps and pressure indicators.
(n)   Hypochlorite flow indication and control.
(o)   Piping and valves and instrumentation for seawater containing up to 2000 ppm sodium hypochlorite.

These equipment requirements demand a high specification to ensure that a sound, maintenance-free design is achieved. This poses two problems, one, whether manufacturers will supply standard components with high specifications to this small market, and two, whether customers will accept the high cost.

A typical process and instrumentation diagram is shown in Fig. 12.3. There are three inlet feed seawater pumps each with its own discharge pressure gauge and non-return valve and isolating valves. This then feeds into a 100 per cent standby seawater strainer system requiring isolation valves, motor control, pressure differential and pressure indication. The strainer output now feeds a common manifold from which eight hypochlorite generators are fed. For each individual chlorinator unit, in this case of 60 kg/h capacity, a flow indicator and isolation valve are required. Product output from the generators feeds into large open-topped hydrogen removal tanks, in this case having a capacity of 30 cubic metres. The piping from the generator requires suitable material to carry seawater with up to 2000 ppm hypochlorite. The hydrogen removal tanks have high and low level switches and level transmitters.

The level transmitter takes a signal and modifies it to control the dosing pumps. By altering the pump flow with control by the level transmitter a constant level in the hydrogen removal tanks is achieved. The low level tank switches protect the pumps should a failure in the level transmitter occur. The high level switches will feed to the control panel and switch off the generators feeding the tank when activated.

Transformer rectifiers provide the required D.C. current to the cells. The associated control panel controls the output chlorination level and also contains all safety interlocks and necessary protection for the generators.

The three dosing pumps in this scheme feed a common manifold and enable either of two injection stages to be fed from any pump. At the injection point, motorized control valves are controlled from the panel and allow the operator to select the required dosing points.

Smaller systems used offshore, or in areas where natural venting would be hazardous, have hydrogen removal tanks fitted with duplicated dilution fans. These fans are sized to dilute the hydrogen/air ratio to 1 per cent by volume, which is a

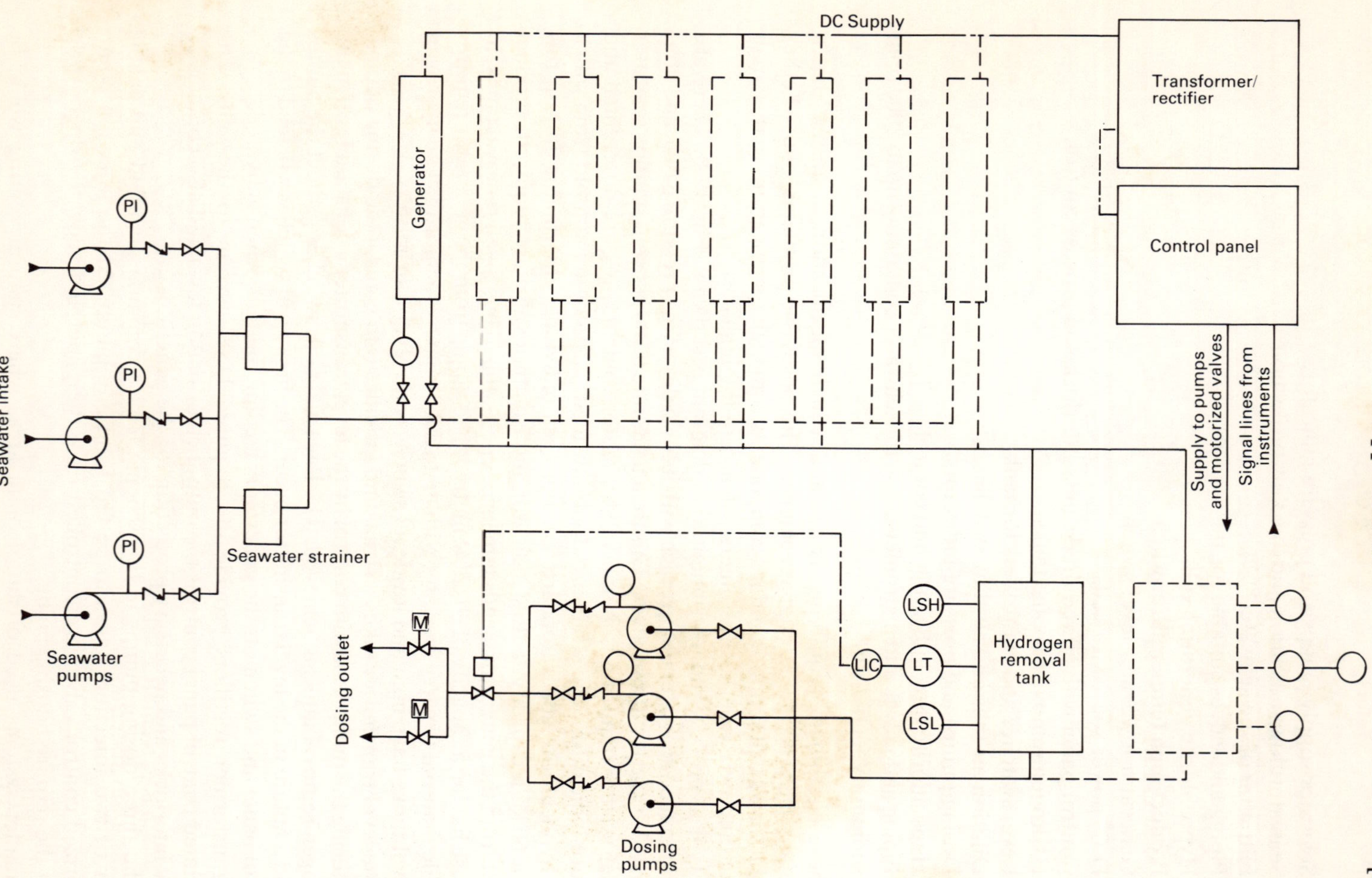

Fig. 12.3 — Electro-chlorination process requirement.

quarter of the lower explosion level for this mixture. Failure of a fan is sensed by a flow switch, this shuts down the generator, starts of the second fan, activates an alarm, and after a predetermined safe venting period shuts down the entire system.

SYSTEM MATERIAL SELECTION

General

The supply of seawater may already be chlorinated at a low level of 0.2 to 1.0 ppm chlorine depending on the take-off position offered by the customer.

Equipment materials will be resistant to corrosion by seawater and are readily available.

The discharge from the hypochlorite generator requires materials with an even higher level of corrosion resistance as in this area there will be a totally different magnitude of chlorine level. Thus, depending on which system is used the concentration of chlorine could be anything up to 2000 ppm in the form of sodium hypochlorite.

Pipework

There is a variety of materials suitable for use with seawater, e.g. types of stainless steel, Monel, bronze, and often Ni-Resist cast iron. Some of these materials are more resistant under continuous flow conditions than stagnant conditions and this may be a consideration in design.

The plastics offer considerable choice, and depending on overall designs for pressure and temperature, can often prove a more cost-effective solution. The seawater–sodium hypochlorite solution is a highly corrosive fluid and, in general, plastics or rubber-lined pipes tend to be used because of their low cost and high corrosion resistance, as follows:

Seawater	Seawater/sodium hypochlorite
Stainless steel	Rubber lined steel
Monel	Saran™ lined steel
Bronze	PTFE lined steel
Cast iron (Ni-resist)	Kynar™ lined steel
PVC	PVC lined GRP
ABS	UPVC, CPVC
	ABS
	Polypropylene

Titanium is extremely corrosion resistant and has been mentioned already as it is one of the main materials for construction of the hypochlorite generators. However, as an alternative, Hastelloy 'C' can be used. Here again the cost is a very strong consideration.

Steel piping has been used with various corrosion resistant linings, rubber, Saran™, PTFE and Kynar™. All types of lined pipe have been used and their use has proved successful. However, with all these forms of linings it is necessary to ensure that the company providing the piping is well proven and has good quality control, since any failure, e.g. chipping or breaking of the lining, would cause rapid corrosion of the steels. Plastics piping, quite naturally has a distinct advantage in cost and resistance. UPVC, CPVC and polypropylene are materials which are well proven and widely used for piping systems. PVC-lined GRP can also give a very good answer where high strength along with good chemical resistance are required. Relative costs for some piping materials in the mid-1980's are shown below:

Material	Cost ratio
Carbon steel, black	1.00
Stainless steel type 304	1.91
Rubber-lined steel	2.15
Polypropylene lined steel, Saran™-lined steel	2.30
PTFE lined steel	2.34
Titanium	7.00
Fibreglass reinforced plastic	1.14
PVC-lined GRP	1.25
UPVC thermoplastic, schedule 80	0.55
CPVC thermoplastic, schedule 80	0.70
Saran™ lined steel	2.00

Valves

The selection of valves for the various applications required for an electrochlorination system reflects the same considerations as for piping although the first priority will be the valve function.

There are four main catagories which are required.
1. On-off service or shut off valves; Gate valves; Plug valves; Ball valves.
2. Throttling service; Globe valves; Butterfly valves; Diaphragm valves.
3. Prevention of back flow; check valves.
4. Miscellaneous: Control valves; Solenoid valves, etc. (these are generally regarded as items of instrumentation).

The most important characteristics of the fluid, which very often limit the choice of valves, are viscosity, corrosion and abrasion, operating pressure, operating temperature and pressure drop.

Valves made entirely from thermoplastic resins, and metallic valves with plastics lined or rubber lined wetted parts are the most suitable for use in seawater/sodium hypochlorite solution. Valves constructed from bronze or stainless steel are very often recommended for use in seawater. Valves constructed of titanium or Hastelloy 'C' are of course most suitable but high costs dictate the use of less expensive but equally suitable materials. A list of available valves is shown below in order of increasing costs:

Material	*Type*
UPVC	ball valves, diaphragm valves, check valves.
Rubber-lined cast iron	diaphragm valves, check valves.
Cast iron, EPDM seat, Teflon™-coated disc	butterfly valves

Rubber-lined cast steel	diaphragm valves
Epoxy resin	butterfly valves.
Kynar™ or Saran™ lined cast iron	diaphragm valves, check valves.
Kynar™ or Saran™ lined cast steel	diaphragm valves, check valves

Diaphragms and seats are constructed of Hypalon™ or rubber suitable for seawater/sodium hypochlorite solution.

As seen from the above list, materials of construction will limit the choice of valves to diaphragm and butterfly valves which are, of course, most suitable for throttling service or flow control. However, although valves suitable for on-off service are not suitable for throttling service the reverse is possible, so diaphragm or butterfly valves will be used as shut-off valves.

Instrumentation, flow and pressure indication
Chlorine residual analysis
Determination of residual chlorine in treated waters can be carried out either by analytical chemical methods or by chlorine residual analysers.

Automatic chlorine residual analysers are available which can be used to control the output of a continuously operating generator. In this case chlorine production and chlorine demand are paced together so that chlorine is not wasted and residual chlorine is kept at acceptable levels.

Flow indication
Flowmeters are provided at the inlets of each generator module, so that total seawater flowmeters are not required. These are available and are either orifice plate type or pitot tube type meters. Flowmeters are offered when a shock chlorination programme is followed so that the volume of solution injected can be monitored.

Pressure indication
Pressure gauges for installation at the outlet of hypchlorite dosing pumps and hypchlorite generators are suitable for sodium hypochlorite solution. These are constructed of stainless steel moving parts with PTFE-lined stainless steel wetted parts and PTFE-lined diaphragms. The gauges are suitably protected for operation in seawater environments. Manufacturers' standard gauges with Monel™ wetted parts are suitable for seawater and are readily available.

Differential pressure switches
These are of similar construction to the pressure gauges already described. They are installed across strainers so that they either initiate strainer self-cleaning operation or signal that the strainer requires maintenance.

Switches for sodium hypochlorite solution have all wetted parts made in plastics material normally Kynar™ or PVC. Pressure switches with all wetted parts in Hastelloy 'C'™ are available but at much higher cost.

Seawater strainers
The seawater supplied to the cells should be strained to remove suspended solids. Electrochlorination systems usually require filtration within the range 500–800 microns.

　　Strainers can be obtained with UPVC, bronze or epoxy-lined cast iron or steel bodies with plastics, stainless steel or titanium screens. In large installations self-cleaning seawater strainers are often required. In smaller installations manually cleaned dual strainers are most suitable. It is obvious that the cost will increase on going from UPVC to bronze to rubber lined cast iron.

Pumps

The materials requirement for seawater pumps is well known and 'off the shelf' pumps are readily available; however, seawater containing up to 2000 ppm of hypochlorite requires further consideration of materials to be used.

　　All-titanium or plastics pumps are available, the former being very expensive, the latter restricted by pressure limitation. Metallic pumps with suitable linings are therefore considered. Here again the rubber, PTFE and other plastics type linings are available but the design curves for the pump must be followed.

　　A high pressure pump run against lower than design head will easily cavitate. The low static pressures developed across the centrifugal open impeller design pump can pull the lining into the impeller and damage the lining. Once damaged the failure is rapid.

Generator cell design consideration

In order to appreciate fully the generation of hypochlorite by the use of an electrochlorinator some basis facts must be considered.

　　For hypochlorite generating cells consisting of an anode and a cathode in seawater the output of chlorine in the form of hypochlorite is proportional, by Faraday's law, to the current passed between these electrodes. It has already been seen that when this current passes chlorine is released from the anode and combines with water to form hypochlorite as hydrogen is released from the cathode. The seawater which carries the current from the anode to the cathode has a resistance related to the resistivity of the seawater and consequently is required to overcome this in order that the current may pass. There is therefore a voltage drop which is equal to the current multiplied by the resistance of the seawater. Unfortunately, there is a natural barrier which prevents current flowing until a certain voltage called the overpotential is overcome. This overpotential depends on the nature of the precious metal coating on the anode and the material of the cathode. A typical curve is shown in Fig. 12.4. The slope of the voltage to current line is dependent on the resistivity of the sea water.

　　For a single monopolar design cell the overpotential voltage for platinum to titanium is 3.3 volts and similarly the overpotential for a typical platinum-irridium coating to titanium is 2.8 volts. As no chlorine in a hypochlorite form is generated until current is passing, it is obvious that the platinum alloy coating produces chlorine at a lower voltage level and therefore requires less power. Therefore, there are two voltages to overcome, the initial overpotential voltage Vo plus the resistive voltage, Vr.

$$V \text{ (Total)} = Vo + Vr$$

$$Vr = I/K$$

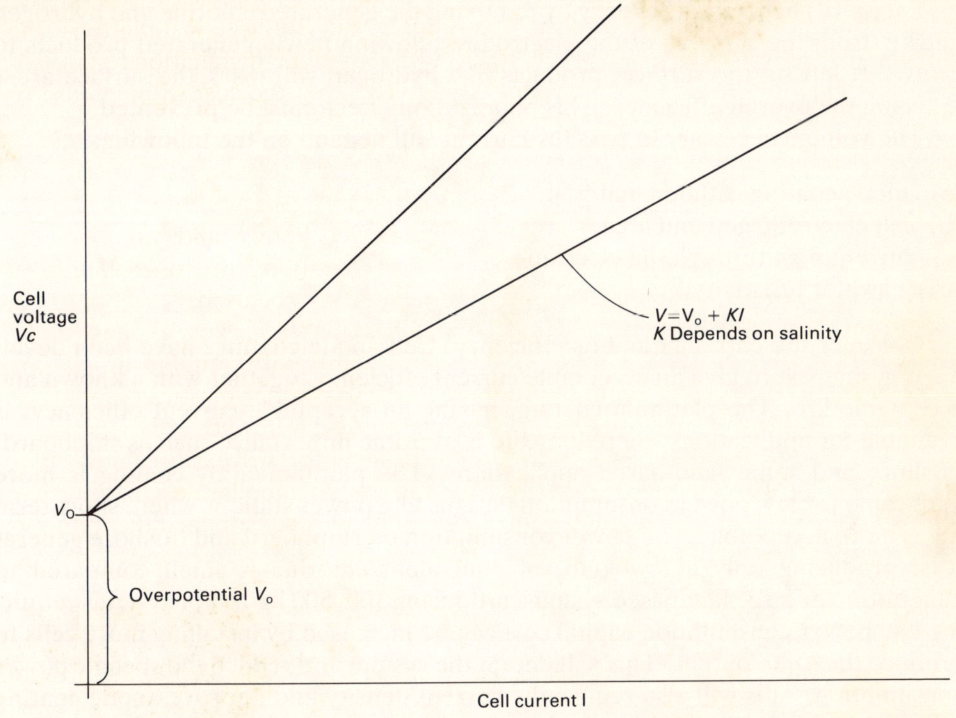

Fig. 12.4 — Electrochemical cell characteristics.

where I is the current passing between the electrodes and K is a constant dependent on cell design.

The resistivity of seawater is measured in ohms-cm and is dependent on the seawater temperature and the total dissolved solids which can carry current in that seawater. The overall cell resistance is a function of this resistivity and the electrode gap between the electrodes and the area through which the current passes. It is also affected by the material resistance of anode and cathode connection resistances, edge losses, stray currents, etc. It therefore follows that the power required to produce a given output is the current multiplied by the voltage and in the case of a cell with a low power consumption, i.e., kW/kg of chlorine, both the current and the voltage must be considered.

The current I passing through the cell will produce an output dependent on the efficiency of the cell to convert NaCl (salt) to usable HOCl (hypochlorite). This efficiency depends on the following factors:

(a) anode coating/cathode material;
(b) seawater temperature;
(c) seawater salinity;
(d) seawater velocity over the electrode surface;
(e) cell physical design, i.e. electrode gap, etc.

Factor (d) helps the efficiency by removing the generated chlorine and hydrogen quickly from the surfaces of the electrodes, allowing newly generated products to evolve. If left on the surface, products like hydrogen will mask the surface areas reducing the overall efficiency. This polarization effect must be prevented.

The voltage necessary to pass the current will depend on the following:

(a) anode coating/cathode material;
(b) cell electrode gap and area;
(c) current flow through the electrodes;
(d) seawater resistivity.

Consider the current I and its efficiency. Cell anode coatings have been developed in the past to give an acceptable current efficiency together with a known and acceptable life. The platinum coating, having an acceptable current efficiency, is available for applications where long life is of prime importance such as shipboard, offshore and some land-based applications. The platinum/alloy coating is more applicable for low power consumption designs like power stations where short-term life is more acceptable. The power consumption on shipboard and offshore generators, producing only a few kg/h, of equivalent chlorine is small compared to generators for large landbased systems producing 100–500 kg/h. For systems requiring low power consumption capital cost can be increased by installing more cells to produce the same output. This will derate the system and reduce the specific power consumption. This will also reduce the current density and improve anode coating consumption (see Fig. 12.5).

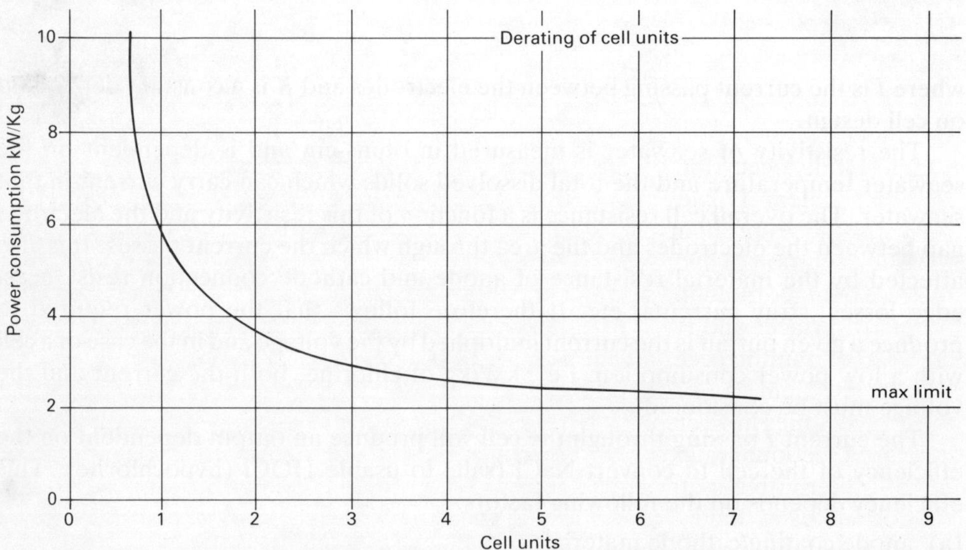

Fig. 12.5 — Electrochlorination system.

In considering the voltage across the cell required to pass the given current it must be remembered that the nature of the anode coating will affect the overpotential component and that the choice of anode coating also depends on other requirements. The cell electrode gap and current flow areas are design parameters which must also be considered. In theory the electrode gap does not affect the current efficiency of the cell but in practice more of the total current passes and better mixing occurs as the anode and cathode surfaces come closer together. As this gap is increased total current passed falls due to losses from the edges of the two electrodes. Therefore, the higher the loss in current passing between the plates the lower the effective current becomes.

Unfortunately both a high velocity over the electrodes and a small gap between the electrodes are required. Both these requirements result in high seawater flow rates and high pressure drops. These two points in particular would demand pumps of high flow rates and large pressure heads. The large pressure requirement may in fact be too great for the cell design itself. The most efficient cell design would require large electrode surface areas, large material thickness for good current distribution and small electrode gaps. This would minimize stray current losses along the edge of the electrode. But this results in electrodes of impractical size and increased cost. A compromise is therefore needed to obtain the most cost-effective design.

Thus, in summary the major points for consideration are:

(a) anode coating/cathode material;
(b) electrode gap;
(c) electrode surface area;
(d) seawater velocity over electrodes;
(e) pressure drop through cells;
(f) overall physical size and strength;
(g) overall operating cost, i.e. capital and running costs.

Hydrogen
This naturally occurring by-product when handled correctly in the product discharge is not a problem. However, on some of the plate type designs of generation cells its presence causes problems.

The evolution of hydrogen on the titanium cathode plates has resulted in titanium hydride being formed in various low flow areas. The titanium hydride remains in the titanium surface, and leads to expansion. This results in an imbalance and stresses across the titanium section and buckling of the plates.

The problem can be overcome by various means. In monopolar plate designs some companies have changed the cathode material for Hastelloy 'C', and some have increased sectional thickness, both actions resulting in higher cost. For bipolar plate designs alternative material and thicker sections have been used. In designs using tubular cell and high flow velocities the circular cross section provides greater strength and the high turbulence quickly removes the hydrogen as it is evolved.

In some flat plate systems the presence of hydrogen is claimed to assist the overall process. Hydrogen, being a very light gas, rises rapidly up and over the electrodes, increasing the turbulence within the cell between the electrode plates. However, a corresponding concern arises when it is remembered that a small quantity of oxygen is released along with the hydrogen. This in itself may be a problem if an ignition source is available but also the rate of precious metal coating depletion will be rapidly increased at the water/gas interface [24].

Coating/plating, wear rate considerations
Plating

Platinum has a small but finite anodic consumption or dissolution rate [24,25]. The normal rate experienced by Electrocatalytic Ltd for platinum coatings in 3 per cent brine at 20°C is in the order of 6–8 mg/ampere year (mg/ay). There are many factors affecting the overall result but consideration should be given to surface roughness, initial strike rate for electrode deposits, surface cleaning, titanium substrate grain size, plating bath salt composition, bath temperature and pH value in the production of the coating. The story does not finish there since the design shape of the platinum anodes and bath, the plating rate, and other parameters all effect the final coating performance and in-service dissolution rate.

Current waveforms
In 1963 Losev [26] published a paper and referred to the effect of current interruptions. It was shown that an interruption of an applied anodic current would cause loss of the protective passive film, resulting in a short period of increased platinum dissolution. Further and later investigation has shown that although in most cases the D.C. current sources have a component of ripple, the frequency is too high to cause effect. However, consideration of the effect on the anodic corrosion rate of platinum may be necessary if pulses in the lower frequencies of 1–2 Hz were applied.

Dilute brine solutions
Work completed on platinum anodic corrosion has indicated an increase in rate when used in very low brine concentration. In low brine concentration as found in estuaries where the evolution in the cell of oxygen and chlorine have taken place simultaneously the corrosion rate increases. The platinum group alloys however show little increase in wear rate at these low brine concentrations [25].

Organics
One further area of consideration is the effect of hydrocarbons on the platinum anodic corrosion rate. The presence of hydrocarbons such as diesel fuel greatly accelerates the anodic corrosion rate of platinum and of ruthenium and iridium oxides.

Water temperature
Platinum anodic corrosion was mentioned above as being in the order of 6–8 mg/ay. This rate varies very little over the normal operating temperature range even when approaching 0°C. However, most of the platinum group alloys, platinum, iridium, ruthenium, etc., have an increased wear rate at temperatures below 5°C. It has been shown by Marshall and Millington [25] that dissolution of platinized titanium

increased with increasing anode potential when the cells were operating in low temperature seawater. It was further explained that the dissolution was primarily caused by the high anode potential and that low temperature was a secondary effect.

Stray current corrosion

In order to reduce stray current corrosion it is advisable to arrange cells in a configuration ensuring negative earth potential at inlet and outlet. This helps to collect all stray charged ions at inlet and outlet. A similar effect can be achieved by placing sacrificial electrodes downstream of the generators. Naturally, the further away from the cells metallic piping can be positioned, the higher becomes the resistance path, and the smaller the level of stray current.

TRANSFORMER/RECTIFIERS AND SYSTEM SAFETY CONTROL

Power supply construction

The material construction standards required to prevent corrosion for the transformer/rectifier equipment supplied are usually determined by customer requirements. The area into which they are installed is specified and will vary from a simple, safe area, indoor, IP21 standard, to a fully certified, explosion proof, unit suitable for ZONE 1 operation in salt-laden atmospheres.

The rectification in standard transformer/rectifiers is either by transductor or thyristor although for special applications and large output generators thyristor rectification is more common. Full wave rectification does offer a relatively smooth D.C. output and results in a 6-pulse output for 3 phase voltage supplied in units. This is acceptable for the precious metal dissolution problems mentioned in the previous section as high frequency ripple is not seen to increase platinum wear rate. However, as these generating systems require a full range of controlled output the thyristors will be phased back and firing very late on the wave rise of the current. This can create high voltage spikes which the transformer/rectifier design must control to prevent coating corrosion of the cell anode. These peaks can be up to five times the rectified level.

Consideration of material constructions within the D.C. power unit depends on the installation environment. A salt-laden circulating air in contact with unprotected hot parts of the unit will have expensive results. Winding classification, thyristor sizing, heat sink material, control card protection, etc., are all important in the final design choice, but they must be cost effective and afford a reliable design.

System control
Current control

It is obvious to most users of electrochlorination systems that good control of the dosing levels is of paramount importance. If there is over-chlorination corrosion rates will be increased; with under-chlorination the marine fouling increases.

A good chlorination system must be adjustable and, once set, must maintain the product dosing levels within acceptable limits. This is normally achieved by having a good current control design such that the set current is maintained during changes of seawater salinity and temperature. This requirement, although costly, is considered to be a fundamental necessity. The Faradaic efficiency will change with salinity and temperature but the change is acceptably small and output is still basically proportional to current.

Voltage control

There are limits to voltages across cell electrodes due to possible corrosion of the cathode. Titanium is widely used by designers due to the protective coating of titanium oxide which forms on the titanium cathode. This protection will however break down when voltages above 8 V are applied. The level of voltage does vary depending on other elements in the water but as some chlorinators may be used in varying water qualities this level is considered a maximum.

The control system will allow the cell voltage to rise as temperature or salinity decreases until the limit discussed above is reached. Further decreases will demand more voltage to maintain the current between the electrodes but the system design will now decrease the current demand to match the voltage limit. This control system does have a further advantage in that cells supplied with very low salinity seawater will have the current reduced. This will help reduce the effect of the high wear rate at low salinities for platinum anodes.

CORROSION OF MATERIALS ASSOCIATED WITH CHLORINATION

The use of metals such as titanium, alloys of copper and nickel, stainless steel, cast steels, etc., has been discussed. Designers must be aware of the effects on the seawater inlet system and of the effect on the customer's system being chlorinated. This is discussed in other pages in this volume but some brief references to work completed on chlorination effect on metals are given below.

Tests completed by Ferrara, Toscherberg; Moran [27] on passive film forming alloys (Ti50, Ti6Al-4V and IN625) conducted in chlorinated seawater (1 mg/l) at flow velocity of 0.3 m/s and 2.4 m/s showed a marked electro-positive shift in open circuit potential whereas tests on non-passive film forms (CA715 and Monel 400) showed a slight negative shift. No corrosion occurred on passive film-forming alloys but non-passive film-forming alloys did show corrosion. Monel (Ni70, Cu30) corroded less in chlorinated seawater whereas CA715 (Cu70, Ni30) showed no significant difference between natural and chlorinated seawater (Table 12.1).

Galvanic couple tests for CA715 and Monel 400 showed significantly lower corrosion rates in chlorinated seawater compared to natural seawater (Table 12.2).

In most cases the corrosion rates were reduced in chlorinated water for passive and non-passive film-forming alloys. However, the corrosion rates increased for increases of velocity of 0.3 to 2.4 m/s, although still less than in natural seawater. For further information refer to [28].

REFERENCES

[1] Mussini, T. and Fiati, L. G. *Encyclopedia of Electrochemistry of the Elements,* Vol. 1 (Ed. A. J. Bard) Marcel Detter, N.Y., 1973).

[2] Kelsall, G. H. 'A Review of Hypochlorite Electrogeneration'. The Electricity Council Research Centre, *ECRC Report/N1059,* June 1977.

Table 12.1 — Average open circuit potentials [Ag-AgCl Ref. Electrode]

Alloy	Average open circuit potential, (V)			
	Natural seawater		1 ppm hypochlorite	
	0.3 m/s	2.4 m/s	0.3 m/s	2.4 m/s
Ti50	0	+0.14	+0.25	+0.45
Ti6Al4V	+0.12	+0.15	+0.15	+0.40
IN625	+0.10	+0.10	+0.35	+0.35
MONEL	+0.09	+0.08	+0.14	+0.12
CA715	−0.18	−0.16	−0.22	−0.22

Test parameters: Duration, 60 days; temperature, 30°C.
Chlorination, 1 mg/l free available chlorine.
Data from [27].

Table 12.2 — Corrosion data for galvanic couples
Average[a] corrosion rate, microns per year

Couple	Area ratio	Anodic Material	Natural seawater		Chlorinated seawater	
			Velocity			
			0.3 m/s	2.4 m/s	0.3 m/s	2.4 m/s
Ti50-CA715	50:1	CA715	467±170	813±300	51±20	340±208
Ti50-Monel	50:1	Monel	315±119	1186±475	23±5	213±178
Ti50-IN625[b]	50:1	NA	Negligible corrosion on either material			
IN625-Monel[c]	2:1	Monel	318±135	330±292	33±2	48±10
Ti50-CA715	1:1	CA715	69±10	107±33	36±15	51±10
Ti6Al4V-CA715	1:1	CA715	99±15	109±41	30±8	43±5
IN625-CA715	1:1	CA715	229±84	267±234	33±5	61±15
Monel-CA715	1:1	CA715	104±38	99±43	53±10	84±0

[a]Average of three tests±standard deviation.
[b]One test.
[c]Average of two tests.
Data from [28].

[3] Pourbaix, M. *Atlas of Electrochemical Equilibria in Aqueous Solutions*, National Association of Corrosion Engineers, 2400 Westloop South, Houston, Texas 77027, USA 1974.

[4] Muller, F. *Z Elektrochem.* **6**, (1899, 8, 11; **8**, 1902, 515,665; **9**, 1903, 171,195. Also Foerster, F. *Trans. Am. Electrochem. Soc.* **46**, 1924, 23 and F. Forester, *Elektrochem. Wasseriger Lozungen* (4th edn.) J. A. Barth, Leipzig, 1923 pp 643, 1923.

[5] Landolt, D. and Ibl, N. *J. Electrochem. Soc.*, **115**, 713, 1968.

[6] Spalding, C. W. *A.I.Ch.E.J.*, **8**, (5), 685, 1962.

[7] Landolt, D. and Ibl, N. *Electrochimica Acta,* **15**, 1165 1970; *J. Applied Electrochem.* 201, 1971.

[8] Jaksic, M. M., Despic, A. R., and Nikolic, B. Z. *Elektrokhimiya,* **8**, (11), 1573, 1972.
Jaksic, M. M. *J. Electrochem Soc.* **121**, (2), 70, 1974.
Jaksic, M. M. *Electrochimica Acta,* **21**, (12), 1127, 1976.
Jaksic, M. M. *J. Appl. Electrochem.,* **3**, 307, 1973.

[9] Ibl, N. and Venlzel, J., Ph.D. Thesis, E.T.H., Zurich Prom. Nr. 3019 (1961), *also* Ibl, N. *Chem. Ing. Techn.* **35**, 353, 1963.

[10] Hammar, L. and Wranglen, G., *Electrochimica Acta,* **9**, (1), 1, 1964.

[11] Beck, T. R. *J. Electrochem. Soc.* 1/969, **116**, (7), 1038, 1969.

[12] Lartey, R. B. Ph.D. thesis, Salford University (1976), 'Hypochlorite Generators'.

[13] Heal, G. R., Kuhn, A. T. and Lartey, R. B. *J. Electrochem Soc.,* **124**, 1690, 1974.

[14] Booth, J. S., Hamzah, H. and Kuhn, A. T. *Electrochimica Acta.,* **25**, (10), 1347, 1980.

[15] Kuhn, A. T. and Hamzah, H. B. *Chemie Ing. Tech.,* **52**, (9) 762–3, 1980.

[16] Booth, J. S. and Kuhn, A. T. *J. Appl. Electrochem.,* **10**, 233, 1980.

[17] Cook, D. and Ormerod, W. G., *C.E.G.B. Report SSD/NE/N14* (Dec. 1966) — (1974).

[18] Houghton, R. W. and Kuhn, A. T. *J. Appl. Electrochem.* **4**, 173, 1974.

[19] Connolly, B. J. *Process Eng.* (Oct. 1973) 96, *idem* Water Services (Oct. 1975).

[20] Kuhn, A. T. and Lartey, R. B. *Chemie-Ing. Techn.* **47**, (4), 129, 1975.

[21] Kuhn, A. T. *Processing,* 1975 (March) 6–7; (April) 10–12.

[22] Anderson, E. P. T. J. Lamb, (Electrocatalytic Ltd. Minerals & Chemicals Corp, 430 Mountain Avenue, Murray Hill,. New Jersey, USA (Brit. Pat. 1396019 (29.5.75).

[23] White, G. C. *Disinfection of Waste Water and Water for Re-Use,* Van Nostrand Reinhold Co., N. Y., USA 1978.

[24] Anderson, D. B. and Vines, R. F. *Anodic Corrosion of Platinum in Seawater,* R. D. Dept, International Nickel Co, Inc N.Y., 1963.

[25] Marshall, C. and Millington, J. P.: 'Platinum loss from platinised titanium in brine cells, E.C.R.C. Capenhurst *J. Appl. Chem.* **19**, (10) 1969.

[26] Losev, V. V.: 'Determination of rate of corrosion of passive metals, *Zh. Priklad, Khim.,* **36** (7) 1963.

[27] Ferrara, R. J. and Taschenberg, L. E. David Taylor Naval Ship Research and Development Centre, Bethesda, Maryland 21218. *Paper No: 211 Corrosion 85,* The Effects of Chlorinated Seawater on Galvanic Corrosion Behaviour of Alloys used in seawater piping systems.

[28] Engelhard Corporation. New Jersey. An overview on the relationship of biological marine fouling and corrosion, for the 23rd Annual Liberty Bell Corrosion Course, 1985.

Index